CLEP

College Level Examination Program

College Mathematics

Andy Gaus, MS
Kathleen Morrieson, MS

XAMonline

Copyright © 2016

All rights reserved. No part of the material protected by this copyright notice may be reproduced or utilized in any form or by any means, electronic or mechanical, including photocopying or recording or by any information storage and retrievable system, without written permission from the copyright holder.

To obtain permission(s) to use the material from this work for any purpose including workshops or seminars, please submit a written request to:

XAMonline, Inc.
21 Orient Avenue
Melrose, MA 02176
Toll Free: 1-800-301-4647
Email: info@xamonline.com
Web: www.xamonline.com
Fax: 1-617-583-5552

Library of Congress Cataloguing-in-Publication Data
Morrison, Kathleen

CLEP College Mathematics/ Kathleen Morrison
ISBN: 978-1-60787-532-1

1. CLEP   2. Study Guides   3. College Mathematics

Disclaimer:

The opinions expressed in this publication are the sole works of XAMonline and were created independently from the College Board, or other testing affiliates. Between the time of publication and printing, specific test standards as well as testing formats and website information may change that are not included in part or in whole within this product. XAMonline develops sample test questions, and they reflect similar content as on real tests; however, they are not former tests. XAMonline assembles content that aligns with test standards but makes no claims nor guarantees candidates a passing score.

© Can Stock Photo Inc./sdmix/5783896

Printed in the United States of America

CLEP College Mathematics

ISBN: 978-1-60787-532-1

Table of Contents

Overview of the CLEP Test 6

Algebra and Functions 24
    Solving Equations and Inequalities. 24
    Linear Equations. 26
    Graphing a Linear Equation 27
    Absolute-Value Equations and Inequalities 31
    Functions 33
    Systems of Linear Equations 40
    Transformations 48

Counting and Probability 61
    Sample Space Size 61
    Probability Calculations 63

Data Analysis and Statistics. 71
    Measuring Characteristics of Data 71
    Displaying Statistical Data 74

Financial Mathematics. 80
    Percents and Percent Change 80
    APR and Interest Rate Calculations 82
    Present and Future Value 83

Geometry. 85
    Polygons: Triangles, Quadrilaterals, and More 85
    Applications of Polygons: Perimeter, Area, Congruence, Similarity,
        and Pythagorean Theorem. 90
    Parallel and Perpendicular Lines: Properties and Applications. 101
    Circles: Properties, Arcs, Angles and Applications 106

Logic and Sets. 111
    Logical Connectives and Quantifiers 111
    Deductive Reasoning and Validity 114
    Set Theory 117

Numbers 119
    Order of Operations 119

Number Sets, Classification, and Theory 121
Real Numbers. 127
Applications of Numbers. 131
Measurement 133

Sample Test 1 139
Answer Key 1 163
Rationales for Test 1 164

Sample Test 2 197
Answer Key 2 220
Rationales for Test 2 221

Sample Test 3 253
Answer Key 3 276
Rationales for Test 3 277

Meet the Authors

Andy Gaus has created math content online and in print for such major publishers and content providers as Glencoe, Holt, Pearson, Aptara Corporation and Brown Educational Network. He is also a theater pianist who has played numerous shows in the Boston area, where he lives.

Kathleen Morrison is a certified Math and Science teacher licensed in the state of Illinois. Her classroom experience spans 16 years covering subject areas from Pre-Algebra to Pre-Calculus, Geometry to Trigonometry, along with interdisciplinary experience combining Algebra and Physical Science. She is currently working as a private mathematics tutor, a substitute teacher, and a writer of mathematics practice exercises.

Overview of the CLEP Test

CLEP is a credit-by-examination program that helps students earn college credit for previous experience or learned material. CLEP offers 33 exams in five subject areas, covering material taught in courses that may generally be taken in the first two years of college. A complete list of exams is available at https://clep.collegeboard.org/exam. Find out if an institution accepts CLEP by visiting the following site: https://clep.collegeboard.org/search/colleges.

Areas Tested in CLEP College Math

The CLEP College Mathematics examination covers material generally taught in a college course for nonmathematics majors and majors in fields not requiring knowledge of advanced mathematics. You may find that you don't know the answer to every single question, as you are not expected to have learned every topic on the test. But, as you prepare for the test, use the topic lists below to study and ensure that you are somewhat knowledgeable about all the topics.

The test consists of 60 multiple-choice questions; a candidate has 90 minutes to complete the test.

Approximately half the examination expects the candidate to solve routine, straightforward problems, while the other half requires an understanding of concepts and skill applications.

A scientific calculator is part of the exam software and is available for use during the entire test. Candidates are strongly encouraged to familiarize themselves with the calculator by downloading a free 30-day trial. http://www.infinitysw.com/ets/

Algebra and Functions 20% (15 questions)

Solving equations and inequalities; graphing linear equations, inequalities, and linear systems; evaluating and interpreting functions; transformations of functions; reflections and symmetry; linear and exponential growth; various applications of algebra.

Counting and Probability 10% (6 questions)
Fundamental counting principle, combinations, permutations, probability of independent, mutually exclusive, complementary and conditional events, expected value, probability applications.

Data Analysis and Statistics 15% (9 questions)
Representation and interpretation of data presented in tables, bar graphs, circle graphs, scatterplots, and histograms; numerical data summary (mean, median, mode, range); standard deviation and normal distribution; applications of data.

Financial Mathematics 20% (12 questions)
Percents and percent change, annual percentage rate (APR), effective interest rate, simple and compound interest, present and future value, mathematical applications of financial concepts.

Geometry and Measurement 10% (6 questions)
Perimeter, area, similarity, triangle and quadrilateral properties, Pythagorean Theorem, parallel and perpendicular lines, properties of circles including circumference, area, sectors, central and inscribed angles, and geometric applications.

Logic and Sets 15% (9 questions)
Logical operations and statements (conditional, conjunction, disjunction, negation, hypothesis, conclusion, inverse, contrapositive); set relationships including Venn diagrams, union, intersection, and subsets; set applications.

Numbers and Operations 10% (6 questions)
Operations, elementary number theory, divisibility, primes and composites, the fundamental theorem of arithmetic, absolute value, real numbers (integers, rational numbers, and irrational numbers), measurement and conversion, precision, and scientific notation.

New to Algebra and Functions

Models of growth exist in many other mathematical patterns. Two more notable growth patterns are as follows:

Quadratic Growth $h(x) = ax^2$

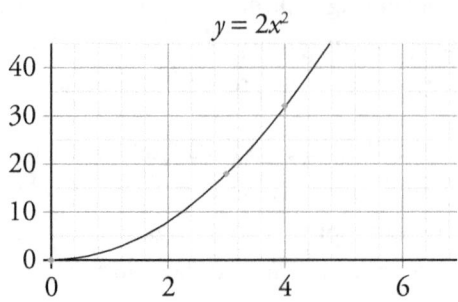

Exponential Growth (with $n > 0$) $f(x) = an^x$

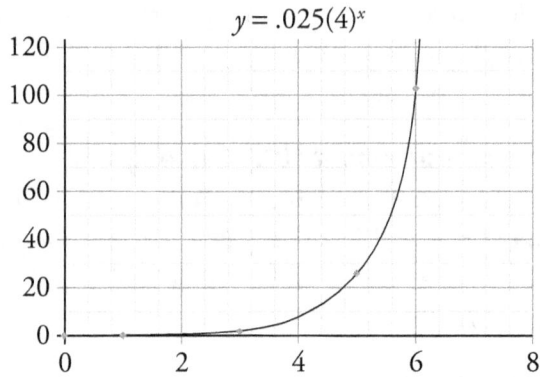

Example: Describe the translation from $y = x$ to $y = x + 7$

The line $y = x$ has a slope of 1 and goes through the origin. The translated line, $y = x + 7$, can be rewritten as $(y - 7) = x$, which predicts a vertical move 7 spaces up, with a new y-intercept at $(0, 7)$. The line will still have a slope of 1.

Example: Reflect the line $y = -2x - 5$ over the x-axis.

A table of values for the original line can be as follows:

x	−4	−2.5	0	1
y	3	0	−5	−7

Since a set of points reflected over the x-axis replaces every y with its opposite, the table of values for the reflection is as follows:

x	−4	−2.5	0	1
y	−3	0	5	7

When graphed, these sets of data create two lines that appear symmetrical with respect to the x-axis.

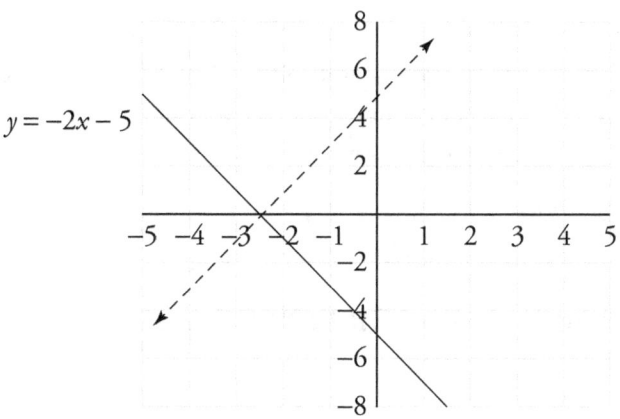

To find the equation of the reflection, take the original equation and replace y with $(-y)$. Then solve for the "new" y.

$y = -2x - 5 \rightarrow -y = -2x - 5$

$y = 2x + 5$ is the equation of the reflection.

New to Counting and Probability

Example: How many different ice cream sundaes can be created when selecting from 3 different flavors of ice cream and 4 different sauce toppings, either with or without nuts? By the fundamental counting principle, the number of possible outcomes is $3 \times 4 \times 2 = 24$.

Example: What is the probability that an even number will be rolled with the toss of a die? Given a standard, 6-sided die, 3 sides, namely 2, 4, and 6, fulfill the desired outcome out of the six sides of the die.

Therefore, $P(\text{even}) = \dfrac{3}{6}$.

Example: Given a jar containing 5 chocolate candies, 4 butterscotch, and 2 peppermint, what is the probability of selecting a peppermint? And of NOT selecting a peppermint?

A total of 11 candies fill the jar with 2 of them peppermint. Therefore $P(p) = \frac{2}{11}$.

When $P(\text{not } p)$, the other 9 candies fulfill this outcome, resulting in a probability of $\frac{9}{11}$. The outcomes of peppermint and "not peppermint" are opposites of each other and are called complementary events.

Note: $P(p) + P(\text{not } p) = \frac{2}{11} + \frac{9}{11} = 1$. The probabilities of complementary events always add to 1.

New to Data and Statistics

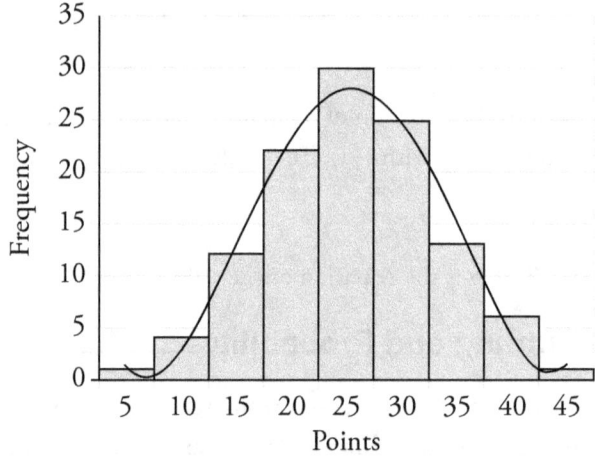

Example: In the graph above, suppose the mean is 23 with a standard deviation of ±4. What values are 3 standard deviations away from the mean?

The standard deviations are measured to the right and left. So 3 standard deviations to the right is $23 + 3(4) = 35$ while 3 standard deviations to the left is at $23 + 3(-4) = 11$

New to Finance

The word "percent" means "per 100" and is a way to quantify sizes and make comparisons between groups of different magnitudes. For instance, if 22% of a group of 100 contestants are to be finalists, 22 people will be chosen. However, 22% of a group of 41 means only 9 will be selected.

Be aware that percent concepts can be presented in various, equivalent formats.

For example, $38\% = 0.38 = \frac{38}{100}$. Cautious observation should be used with decimal places; for instance $0.01 = 1\%$ but $0.01\% = .0001$. Basic percent calculations can be performed as proportions or linear equations.

Example: Proportion

5 is what percent of 20?

The structure of the proportion is $\frac{\text{part}}{\text{whole}} = \frac{p}{100}$.

$$\frac{5}{20} = \frac{p}{100}$$
$$20p = 500$$
$$p = 25$$

The answer is 5 is 25% of 20.

Example: Linear Equation

There are 64 dogs in the kennel. 48 are collies. What percent are collies?

Restate the problem. 48 is what percent of 64?
Write an equation. $48 = n \times 64$

Solve.
$$\frac{48}{64} = n$$
$$n = \frac{3}{4} = 0.75 = 75\%$$

75% of the dogs are collies.

Note that in the proportion method, the percent value is used directly, while in the linear equation method, the percent value must be converted to its decimal equivalent.

Example: The auditorium was filled to 90% capacity. There were 558 seats occupied. What is the capacity of the auditorium?

Restate the problem.   90% of what number is 558?
Write an equation. $0.9n = 558$

Solve. $$n = \frac{558}{.9}$$
$n = 620$

The capacity of the auditorium is 620 people.

Example: A pair of shoes costs $42.00. Sales tax is 6%. What is the total cost of the shoes?

Restate the problem.   What is 6% of 42?
Write an equation. $n = 0.06 \times 42$
Solve. $n = 2.52$

Add the sales tax to the cost. $42.00 + $2.52 = $44.52

The total cost of the shoes, including sales tax, is $44.52.

Example: A share of stock in Wonder Widgets cost $52 at the start of the day and had risen to $83 at the close of trading. What was the percent change in the share price?

First find the actual change in the share price: $83 - 52 = \$31$ change in share price.

When considering percent change, the change should be compared to the original. In this case, consider "31 is what percent of 52?" as 52 represents the original share price.

Solving by Linear Equation
$31 = r \times 52$

$$r = \frac{31}{52} \approx .60 = 60\%$$

Example: A sweater priced at $47 is marked down 15%. What is the new price, to the nearest dollar, of the sweater?

Find:

15% of 47 = d, where d represents by how many dollars the price is lowered.

0.15 (47) = 7.05 dollars decrease, thus the new price of the sweater will be 47 − 7.05 ≈ $40.

APR and Interest Rate Calculations

The financial world relies heavily on the percent calculation, and numerous instances involve calculations that are more complicated, and often more repetitive, than the basic examples shown above.

Often, for instance, a bank will charge an annual percentage rate, or APR, of p% on a loan. In the most basic sense, if a consumer borrowed $100 for a year with an APR of 9%, he or she would owe the bank $109 upon repayment. But this assumes the interest is charged annually, or only one time during the year.

In reality, lending institutions calculate interest at intervals throughout the year. Interest can be charged, for example, quarterly (4 times a year), or daily (365 times a year), or even continuously. This periodic rate affects the actual interest paid on the loan (or, conversely, earned on an investment). A valuable calculation used to understand the reality of interest rate charges is the effective interest rate formula:

$$E = \left(1+\frac{r}{n}\right)^n - 1$$ where r represents the APR, and n is the number of times the interest is calculated in a year.

Example: A credit card offers an APR of 18%. Charges are calculated monthly. Find the effective interest rate on the card.

$$E = \left(1+\frac{r}{n}\right)^n - 1 = \left(1+\frac{0.18}{12}\right)^{12} - 1 \approx 0.196 = 19.6\%$$

This calculation reminds the consumer that interest charges are, in fact, higher than the straight 18% offered with the card.

The effective interest rate formula can be expanded, so to speak, to become the compound interest formula:

$$A = P\left(1+\frac{r}{n}\right)^{nt}$$

This calculation is most commonly used for investments lasting a year or more when interest is awarded at multiple intervals throughout the year. P stands for the principal, or the initial investment amount. The values for r and n are the same as in the effective interest formula, and t represents the number of years that the money is invested. The calculation yields the final balance A.

Example: Suppose a person invests $2,000 in an account earning an annual interest rate of 3.3% compounded quarterly. How much money will be in the account at the end of 5 years?

$$A = 2{,}000\left(1+\frac{0.033}{4}\right)^{4 \cdot 5} = 2{,}000(1.00825)^{20} = \$2{,}357.19$$

As mentioned earlier, a periodic interest rate can be measured in various intervals from months, to days, to seconds, and even to infinitely small intervals, which results in continuous compounding. To calculate interest compounded continuously, a new formula must be used: $A = Pe^{rt}$, where e is the constant base of the natural log function (e is an irrational number and has an approximate value of 2.718).

Example: Find the balance on an account starting with an initial deposit of $750 after 10 years of interest compounded continuously at an annual rate of 4.9%. Using the continuously compounded interest formula: $A = Pe^{rt} = 750e^{.049(10)} = \$1{,}224.24$

Present and Future Value

The concept of present and future value is used to present the idea that money may be worth more (or less) later, when considering its investment potential, or that expenses may be greater as time goes by due to inflation. For instance, in the previous example, the approximate future value of the $750 (present value) under the given investment conditions is $1,224.

Example: Find the present value needed to have, in 18 years, a future value of $25,000 with an investment rate of 5% compounded monthly.

To determine this amount, the compound interest formula can be used with the future value representing A:

$$25{,}000 = P\left(1 + \frac{0.05}{12}\right)^{12(18)}$$
$$25{,}000 = P$$
$$P \approx \$10{,}183$$

In other words, a present value investment of a little over $10,000 can more than double to $25,000 in 18 years, given these investment conditions.

Example: If the rate of inflation averages 2.4%, what will be the future cost, in 5 years, of a gallon of milk with a present value of $3.50? Inflation calculations are performed with the continuously compounding formula.

$$A = Pe^{rt} = 3.5e^{(.024)(5)} = \$3.95$$

New to Geometry

Additionally, an equilateral triangle has 3 congruent angles, and an isosceles triangle has two congruent base angles. (The congruent angles sit at each end of the non-congruent side of the triangle.)

Example: Given equilateral triangle TRY, find $m\angle R$ and length of side RY.

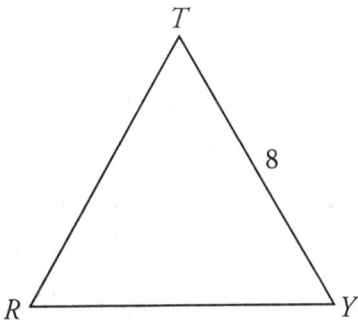

Since the triangle is equilateral, every side is the same length. So RY must also be 8 units long. Also, an equilateral triangle has 3 congruent angles. If the sum of the angles of a triangle is 180°, then each angle measures 180 ÷ 3 = 60. Therefore, $m\angle R = 60°$.

Example: In the diagram below, ℓ_1 is parallel to ℓ_2 and $m\angle a = 55°$. Find $m\angle b$ and $m\angle c$.

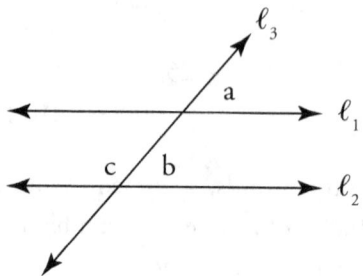

With ℓ_3 as the transversal through 2 parallel lines, $\angle a$ and $\angle b$ are congruent corresponding angles. Therefore, $m\angle b = 55°$. Since $\angle c$ and $\angle b$ form a linear pair, they are supplementary: their measures add to 180°. $180 - 55 = 125 = m\angle c$.

Example: Given circle B, and that the measure of minor arc XY is 100°, find $m\angle B$ and $m\angle P$.

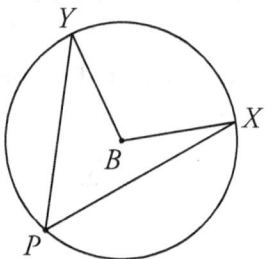

Since $\angle B$ <B is a central angle, its measure is equivalent to the intercepted arc: 100°.

However, $\angle P$, with vertex on the edge of the circle, is an inscribed angle, so its measure is half of the arc: 50°.

The converse of a conditional requires the conclusion and hypothesis to trade places in the statement. For instance: the converse of the statement "If it is snowing outside, then it is cold" is "If it is cold outside, then it is snowing." Note by this example that not every converse created from a conditional will be true. If the converse is true, the conditional is said to be reversible.

The inverse of a conditional is the negation of both the hypothesis and conclusion. Again, given "If it is snowing outside, then it is cold," the inverse is "If it is not snowing outside, then it is not cold."

New to Logic and Sets

Set Theory

Consider a Venn diagram with a finite set of elements, such as letters.

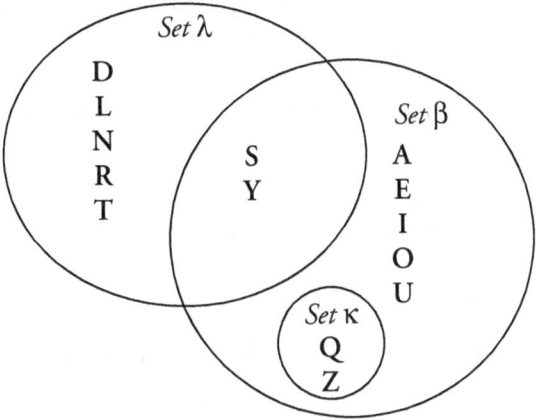

Various symbols and statements can be used to describe the sets and their content in a sort of mathematical shorthand.

The Venn diagram shows *Set β*, *Set λ*. and *Set κ*
- λ = {D, L, N, R, T, S, Y)
- β = {A, E, I, O, U, S, Y, Q, Z}
- κ = {Q, Z}

R is an element of Set λ, but not Set β.
- R ∈ λ
- R ∉ β

The elements Y and S are members of both sets λ and β. This is usually described as the intersection of the two sets.
- λ ∩ β = {S, Y}

Furthermore, as set κ is completely inside set β, it is classified as more than just an intersection. The set κ is a subset of β.
- κ ⊆ β

Two or more sets can be combined by joining them in a union.
- λ ∪ κ = {D, L, N, R, T, S, Y, Q, Z}
- β ∪ κ = β (note: a union does not duplicate elements)

Another operation between two sets is the Cartesian product. The product "λ × κ" (read "cross") is the set of all ordered pairs (*b*, *c*) where *b* is a member of λ and *c* is a member of κ.
- λ × κ = { (D, Q), (L, Q), (N, Q), (R, Q), (T, Q), (S,Q), (Y, Q), (D, Z), (L, Z), (N, Z), (R, Z), (T, Z), (S, Z), (Y, Z) }

Often it is impractical to list every element in a set. Some of these cases can be as follows:
- Ω = {ℜ} denotes the set of all real numbers
- while the set φ = {x ∈ ℜ | x < 2} contains "all real numbers x *such that x is less than 2.*"
- {n | n > 0} describes a set of numbers greater than zero. Note that this set does not have a name, which can often occur.

Example: Find the prime factorization of 68.

Create a factor tree. Keep dividing each branch until a prime number is reached.

This technique shows that $68 = 2 \times 2 \times 17$ or $2^2 \times 17$.

New to Numbers

Calculating within number systems

The absolute value of a number is the distance of the number from zero on the number line. "The absolute value of *n*" is written as $|n|$.

$$|-122| = 122,$$
$$|8.7| = 8.7,$$
$$|3-10| = |-7| = 7$$

Examples: Operations with integers. While the basics of integer computation can be reasoned on a number line, it is essential to quickly recognize the patterns of basic operations. The addition of two negative numbers results in a negative number.

Example: $-3 + -8 = -11$

When adding a positive and a negative number, the number with the greater absolute value dominates the sign of the answer. The value is the difference between the absolute values.

Example: $-6 + 14 = 8$ (positive answer, since $14 > 6$)

Example: $-9 + 5 = -4$ (negative answer, since $9 > 5$)

When performing subtraction on integers, the operation can be written as addition of the opposite.

Example: $7 - (-5)$ is rewritten as $7 + 5 = 12$

When needing to determine the sign of an answer in a multiplication or division problem, simply count up the number of negatives. An even number of negatives will result in a positive product or quotient, an odd number will be negative.

Example: $\frac{-8(-5)}{-4} = -10$ (negative answer from an odd number of negatives)

Operations with rational numbers. Division of fractions should be rewritten as multiplication of the reciprocal. $\frac{3}{7} \cdot \frac{1}{3} = \frac{3}{7} \times \frac{3}{1} = \frac{9}{7}$

Example: The example above shows that radicals can be separated (or joined) by multiplication. This is not the case for addition or subtraction. That is, $\sqrt{7} \neq \sqrt{5} + \sqrt{2}$. We can only add or subtract radicals that have the same index and the *same radicand*.

Example: $2\sqrt{5} + 3\sqrt{5} = 5\sqrt{5}$

Example: $5\sqrt[3]{2} - 3\sqrt[3]{2} = 2\sqrt[3]{2}$

If the radicand is raised to a power different from the index, convert the radical to its exponential form and apply laws of exponents.

Example: $\sqrt[6]{9} = \sqrt[6]{3^2} = 3^{2/6} = 3^{1/3} = \sqrt[3]{3}$

Measurement is a way of quantifying characteristics of physical matter, time, size, and space. Each characteristic has its own basis for measurement. Measuring length, for instance, is distinctly separate from measuring time. When taking measurements, the precision depends on the circumstances and relative size. The distance between two towns does not need to be measured to the nearest thousandth of a mile. The nearest whole mile or tenth of a mile is usually sufficient. Additionally, the measurement tool being used influences the amount of precision expected. Most rulers measure to the nearest millimeter and bathroom scales to the nearest pound or tenth of a pound. Therefore, if any calculations need to be performed with measurements, consider such precision when presenting the answer. For instance, a velocity calculated from data in whole feet and seconds should be rounded to the nearest whole or tenth of a number. Any further decimal places would be unreliable.

Often, the need arises to convert between units of measure. When calculating such changes, use the "unit cancelation method" to make appropriate multiplication or division choices. For instance, since there are 36 inches in a yard, the conversion value can be expressed as

$$\frac{36 \text{ in.}}{1 \text{ yd}} \text{ or } \frac{1 \text{ yd}}{36 \text{ in.}}$$

as needed. Suppose the length of 48 inches needs to be converted into yards. Then inches need to be canceled out while yards need to be part of the answer. Therefore consider:

$$\frac{48 \text{ in.}}{1} \times \frac{1 \text{ yd}}{36 \text{ in.}} = \frac{48}{36}\left(\frac{\text{in.} \times \text{yd}}{\text{in.}}\right) = \frac{4}{3}(\text{yd}) = 1\frac{1}{3} \text{ yd}$$

The unit cancelation method can be used to convert any measured property, and between all systems of measurement.

When calculating with numbers in scientific notation, simply follow the rules of exponents.

Example: $(2.3 \times 10^8)(2 \times 10^{-3}) = 4.6 \times 10^5$ (multiply like bases, add exponents)

Example: $(8 \times 10^4) \div (2 \times 10^{12}) = 4 \times 10^{-8}$ (divide like bases, subtract exponents)

Example: $(3 \times 10^4)^5 = 243 \times 10^{20} = 2.43 \times 10^{22}$ (raise a power to a power, multiply exponents)

Example: What is the probability that an even number will be rolled with the toss of a die?

Given a standard, 6-sided die, 3 sides, namely 2, 4, and 6, fulfill the desired outcome out of the six sides of the die. Therefore, $P(\text{even}) = \dfrac{3}{6}$

Example: Given a jar containing 5 chocolate candies, 4 butterscotch, and 2 peppermint, what is the probability of selecting a peppermint? And of NOT selecting a peppermint?

A total of 11 candies fill the jar with 2 of them peppermint. Therefore $P(p) = \dfrac{2}{11}$. When considering $P(\text{not } p)$, the other 9 candies fulfill this outcome, resulting in a probability of $\dfrac{9}{11}$. The outcomes of peppermint and "not peppermint" are opposites of each other and are called complementary events. Note that $P(p) + P(\text{not } p) = \dfrac{2}{11} + \dfrac{9}{11} = 1$. The probabilities of complementary events will always add to 1.

Why Take CLEP College Math Test?

The CLEP College Math test is an opportunity to earn qualifying scores leading to possible college credit.

How Is the CLEP College Math Test Scored?

One point is earned for each correct answer on a CLEP test. No points are deducted for wrong answers so you should try to answer every question, guessing if needed. Visit the following site for information on the scoring of CLEP tests: http://media.collegeboard.com/digitalServices/pdf/clep/CLEP_scores.pdf

How Is the CLEP College Math Exam Administered?

CLEP exams are administered throughout the year at over 1,900 colleges and universities. Search locations, dates, and registration procedures at https://clep.collegeboard.org/started

Accommodations for Students with Disabilities

If you have a disability, such as a learning or physical disability, which would prevent you from taking a CLEP exam under standard conditions, you may request accommodations at your preferred test center. Contact your preferred test center well in advance of the test date to make the necessary arrangements and to find out its deadline for submission of documentation and approval of accommodations. Accommodations that can be arranged directly with test centers include:
- ZoomText (screen magnification)
- Modifiable screen colors
- Use of a reader, amanuensis, or a sign language interpreter
- Extended time
- Untimed rest breaks

If the above accommodations do not meet your needs, contact CLEP Services for information about other nonstandard options at clep@info.collegeboard.org or 800-257-9558 before you register through My Account.

Getting Ready for Test Day

Be prepared to bring the following at the test center:
- A valid registration ticket from My Account registration portal
- Registration forms or printouts required by the test center. Make sure you have filled out all necessary paperwork in advance of your testing date.
- Administration fees: Bring the administration fee to the test center. Each test center charges an additional fee and sets its own policy for payment.
- Identification: Your driver's license, passport or other government-issued identification that includes your photograph and signature.

You will be asked to show this identification to be admitted to the testing area. The last name on your ID must match the name on your registration ticket.

Testing Tips

1. **Get smart, play dumb.** Sometimes a question is just a question. No one is out to trick you, so don't assume that the test writer is looking for something other than what was asked. Stick to the question as written and don't overanalyze.
2. **Do a double-take.** Read test questions and answer choices at least twice because it's easy to miss something, to transpose a word or some letters. If you have no idea what the correct answer is, skip it and come back later if there's time.
3. **Turn it on its ear.** The syntax of a question can often provide a clue, so make things interesting and turn the question into a statement to see if it changes the meaning or relates better (or worse) to the answer choices.
4. **Get out your magnifying glass.** Look for hidden clues in the questions, because it's difficult to write a multiple-choice question without giving away part of the answer in the options presented. In most questions you can readily eliminate one or two potential answers, increasing your chances of answering correctly to 50/50, which will help out if you've skipped a question and gone back to it (see tip #2). So, read the question carefully.
5. **Call it intuition.** Often your first instinct is correct. If you've been studying the content you've likely absorbed something and have subconsciously retained the knowledge. On questions you're not sure about trust your instincts, because a first impression is usually correct.
6. **Become a clock-watcher.** You have a set amount of time to answer the questions. Don't get bogged down laboring over a question you're not sure about when there are ten others you could answer more readily.

Algebra and Functions

Solving Equations and Inequalities

Both equations and inequalities relate two quantities which may be expressed as any combination of constants, variables, and functions. In an equation, the two quantities are stated to be equal. In an inequality, one quantity is or may be greater than the other.

Equations

An **equation** consists of two expressions linked by an equal sign (statement H1) = (statement H2)

Left Hand Side (LHS) = Right Hand Side (RHS).

If substituting a value for the variable results in LHS = RHS, or a true statement, then the value is a solution for that equation.

Example: $2x = 6$
(LHS) (RHS)

If we substitute 3 for x, we get $2 \cdot 3 = 6$ (True). Therefore, 3 is a solution for the equation.

Example: Is 2 a solution of $2x - 6 = 6x + 1$?

Substituting 2 for x, we get

$$2(2) - 6 = 6(2) + 1$$
$$4 - 6 = 12 + 1$$
$$-2 = 13 \text{ (False)}$$

Therefore, 2 is not a solution.

Inequalities

An **inequality** has the same form as an equation, but the equals sign is replaced by one of the following inequality signs:

   < (less than)
   > (greater than)
   ≤ (less than or equal to)
   ≥ (greater than or equal to)

Example: $x + 2 < 7$ The solution is $x < 5$, meaning that any number less than 5 is a solution of the inequality.

Important Facts About Inequalities

1. *Sense of an inequality:* This is the direction of the inequality. The larger number is always facing the open side.

    Example: $25 > 3$ (greater than)

    Example: $3 < 25$ (less than)

2. *Notation:*

    $\geq \equiv$ "Greater than *or* equal to".

    $\leq \equiv$ "Less than *or* equal to".

    These relations are satisfied if either half of the relation is satisfied.

    Example: $25 \geq 3$ is true if $25 > 3$ is true or if $25 = 3$ is true. Since $25 > 3$ is true, $25 \geq 3$ is true, even though $25 = 3$ is false.

    Example: $0 \leq 0$ is true if $0 < 0$ is true or $0 = 0$ is true. Since $0 = 0$ is true, $0 \leq 0$ is true, even though $0 < 0$ is false.

3. Multiplying or dividing by a negative number *changes the direction* of the inequality.

    Example: $-3x > 6$ Dividing both sides by -3, we get $x < -2$ (note the change in direction)

Properties of Equations and Inequalities

1. We can add any real number to, or subtract any real number from, both sides of the equation (or inequality).

    Example: $3 = 3 \Rightarrow 3 + 2 = 3 + 2 \Rightarrow 5 = 5$ (still true)

    Example: $9 = 9 \Rightarrow 9 - 3 = 9 - 3 \Rightarrow 6 = 6$ (still true)

    Example: $x + 3 = 6 \Rightarrow x + 3 - 3 = 6 - 3 \Rightarrow x = 3$

2. We can multiply or divide both sides of an equation or an inequality by any real number except 0.

Recall: When multiplying or dividing by a negative number we change the direction of the inequality.

Example: $3 = 3 \to 3 \times 2 = 3 \times 2 \to 6 = 6$ (still true)

Example: $8 = 8 \to \dfrac{8}{2} = \dfrac{8}{2} \to 4 = 4$ (still true)

Example: $-2x = 6 \to \dfrac{-2x}{-2} = \dfrac{6}{-2} \to x = -3$

Example: $6 > 2 \to 6 \times 2 > 2 \times 2 \to 12 > 4$ (still true)

Example: $-2 < 6 \to \dfrac{-2}{-2} > \dfrac{6}{-2} \to 1 > -3$ (still true, but with reversed inequality)

Example: $-3x \geq 5 \to \dfrac{-3x}{-3} \leq \dfrac{5}{-3} \to x \leq -\dfrac{5}{3}$ (note reversed inequality)

Linear Equations

A **linear equation** is one in which no variable has a higher power than 1.

Solving Linear Equations

1. Expand to eliminate all parentheses.
2. If there are fractions, multiply each term by the LCD to eliminate all denominators.
3. Combine terms on each side when possible.
4. Perform operations on both sides of the equation to isolate all variables on one side and all constants on the other side.

Example: Solve for x: $3(x + 3) = -2x + 4$

$3x + 9 = -2x + 4$ Expand parentheses.
$3x = -2x - 5$ Subtract 9 from both sides.
$5x = -5$ Add $2x$ to both sides.
$x = -1$ Divide both sides by 5.

Example: Solve for x: $2x + 9 - 3x + 10 = 3x + x - 6$

$-x + 19 = 4x - 6$	Combine similar terms on each side.
$-x = 4x - 25$	Subtract 19 from both sides.
$-5x = -25$	Subtract $4x$ from both sides.
$x = 5$	Divide both sides by -5.

Example: Solve for x: $3x - \dfrac{2}{3} = \dfrac{5x}{2} + 2$

$18x - 4 = 15x + 12$	Multiply each term by 6, the LCD of 2 and 3.
$18x = 15x + 16$	Add 4 to each side.
$3x = 16$	Subtract $15x$ from each side.
$x = \dfrac{16}{3}$	Divide each side by 3.

Graphing a Linear Equation

The graph of a linear equation represents a straight line. It takes two points to define a unique straight line.

1. Choose only 3 values of x.
2. Substitute each chosen value of x in the equation to find the corresponding y-value.
3. Plot the 3 points and join them with a straight line.

Note: It is typically helpful to choose the x-intercept and the y-intercept as the two key points (when possible).

Recall:
- The **x-intercept** is the point where the line intersects the x-axis. To find this point we substitute 0 for y and solve for x.
- The **y-intercept** is the point where the line intersects the y-axis. To find this point we substitute 0 for x and solve for y.

Example: sketch the graph of the line represented by $2x + 3y = 6$

$$\text{Let } x = 0 \Rightarrow 2(0) + 3y = 6$$
$$\Rightarrow 3y = 6$$
$$\Rightarrow y = 2$$
$$\Rightarrow (0, 2) \text{ is the } y\text{-intercept.}$$

Let $y = 0$ $\Rightarrow 2x + 3(0) = 6$
$\Rightarrow 2x = 6$
$\Rightarrow x = 3$
$\Rightarrow (3, 0)$ is the x-intercept.

Let $x = 1$ $\Rightarrow 2(1) + 3y = 6$
$\Rightarrow 2 + 3y = 6$ (subtract 2 from both sides)
$\Rightarrow 3y = 4$ (Divide both sides by 3)
$\Rightarrow y = \frac{4}{3}$

$\Rightarrow \left(1, \frac{4}{3}\right)$ is the third point.

Plotting the 3 points or the coordinate system, we get the following graph:

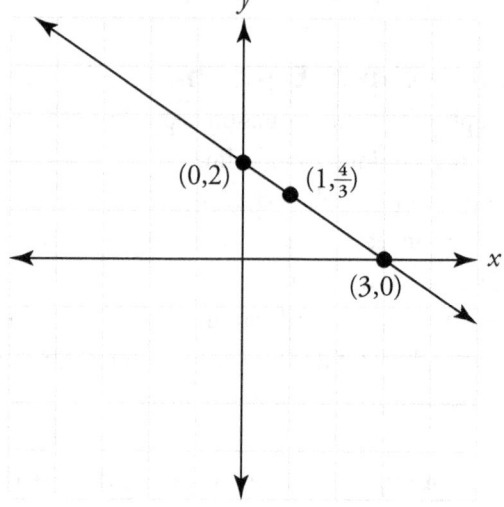

Note: Two points are sufficient to graph the line; the third point is for checking purposes.

Solving Linear Inequalities

Linear inequalities are inequalities in which no variable has a higher power than one.

We use the same procedure used for solving linear equations, but the answer is represented in graphical form on the number line or in interval form.

Example: Solve the inequality, show its solution using interval form, and graph the solution on the number line.

$\frac{5x}{8} + 3 \geq 2x - 5$ Multiply each term by 8 to clear denominator.

$5x + 24 \geq 16x - 40$ Subtract 24 from each side.

$5x \geq 16x - 64$ Subtract 16x from each side.

$-11x \geq -64$ Divide each side by -11; reverse inequality sign.

$x \leq 5\frac{9}{11}$

Solution in interval form: $(-\infty, 5\frac{9}{11}]$ (Note that "[" means $5\frac{9}{11}$ is included in the solution.)

Graph of solution:

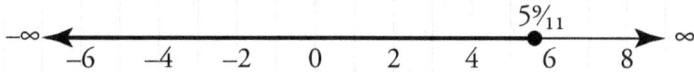

Interval and Graph Notation for Inequalities

- [and] mean that the lower and upper limit, respectively, are included as solutions. In graphing, a *closed dot* (•) indicates the same thing. Inclusive limits are specified with "greater than or equal to" or "less than or equal to" inequalities.
- (and) means that the lower and upper limit, respectively, are excluded as solutions. In graphing, *an open* dot (○) indicates the same thing. Exclusive limits are specified with "greater than" or "less than" inequalities.

Example: Solve the following inequality and express your answer in both interval and graphical form.

$3x - 8 < 2(3x - 1)$
$3x - 8 < 6x - 2$ Distributive property.
$3x < 6x + 6$ Add 8 to each side.
$-3x < 6$ Subtract 6x from each side.
$x > -2$ Divide each side by -3; reverse inequality.

In graphical form:

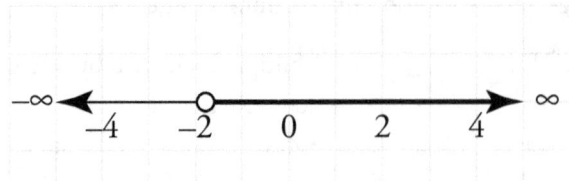

Interval form: (–2,∞)

Note: that the "(" means that –2 is NOT included as a solution.)

Example: Is –2 one of the solutions of the following inequality?

$2x - 6 \leq x + 4$

Substituting –2 for x, we get

$2(-2) - 6 \leq -2 + 4$
$-10 \leq 2$

This is a true statement; therefore, –2 is a solution of the inequality.

Example: Is 3 one of the solutions of the following inequality?

$3x \leq 3 + 2$

Substituting 3 for x, we get $3(3) \leq 3 + 2$, which equals $9 \leq 5$

This statement is false; therefore, 3 is not a solution of the inequality.

Note: A linear equation has one solution, no solution, or an infinite number of solutions. However, a linear inequality can have any number of solutions.

Graphing a Linear Inequality in Two Dimensions

A linear inequality in two variables is similar in form to a linear equation *except* that the = sign is replaced by an inequality sign of >, <, ≥, or ≤. The procedure to graph it is as follows:

1. Graph the equivalent equation with the inequality sign replaced by an equals sign. Use a solid line for this line if the inequality contains the equals sign ($\leq$ or $\geq$); use a dashed line if the inequality contains no equals sign ($<$ or $>$).
2. Pick a point on either side of the line and test whether its x- and y-values satisfy the inequality. If so, mark that region as a solution set with shading or slanted lines. If not, shade the opposite region.

Example: Identify the region that satisfies $3x + 5y < 15$.

1. We graph the equivalent equation of $3x + 5y = 15$ using a dashed line. Substituting $y = 0$ produces an x-intercept of $(5, 0)$; substituting $x = 0$ produces a y-intercept of $(0,3)$.

2. Pick a test point on either side of this line. Pick the origin for simplicity $(0, 0)$. Substitute $x = 0$ and $y = 0$ into the inequality and test:

   $0 < 15$ is true, so accept the region containing $(0, 0)$ and shade it.

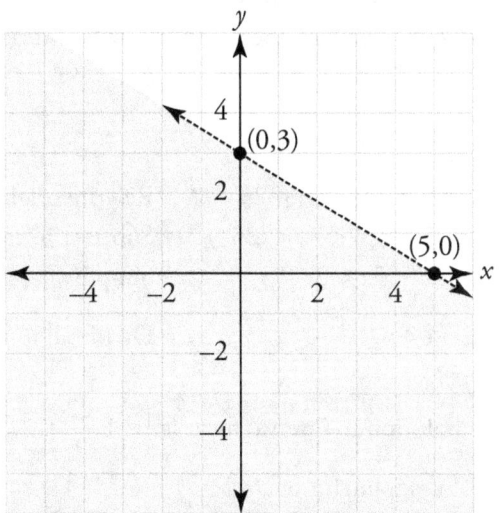

Absolute-Value Equations and Inequalities

An **absolute-value equation** or inequality is one in which one or more of the variables is given as an absolute value. Solving absolute-value equations and inequalities introduces a new level of complexity, because in many cases it is necessary to consider two different values of an expression whose absolute value would be the same and to solve separately for each one.

Absolute-Value Equations

If a and b are real numbers, and k is a nonnegative real number, the solution of $|ax + b| = k$ **is** $ax + b = k$ **or** $ax + b = -k$

Example: Solve for x: $|2x + 3| = 9$

$$2x + 3 = 9 \quad \text{or} \quad 2x + 3 = -9$$
$$2x + 3 - 3 = 9 - 3 \quad\quad 2x + 3 - 3 = -9 - 3$$
$$2x = 6 \quad\quad 2x = -12$$
$$x = 3 \quad\quad x = -6$$

Therefore, the solution is $x = \{3, -6\}$

Example: Solve for x: $|3x - 1| = -3$

Since -3 is a negative number, and an absolute value cannot be negative, there is no solution.

Absolute-Value Inequalities

If a and b are real numbers and k is a nonnegative real number, the solution of $|ax + b| < k$ is $-k < ax + b < k$

Example: Solve $|7x + 3| < 25$

$-25 < (7x + 3) < 25$ Rewrite original inequality.
$(-25 - 3) < 7x < (25 - 3)$ Subtract 3 from each term.
$-28 < 7x < 22$ Simplify.
$-4 < x < \dfrac{22}{7}$ Divide all terms by 7.

Solution in interval form is $\left(-4, \dfrac{22}{7}\right)$.

In graphic form:

If a and b are real numbers and k is a nonnegative real number, the solution of $|ax + b| > k$ is $ax + b > k$ or $ax + b < -k$

Example: Solve $|2x - 7| > 5$

$2x - 7 > 5$ $\qquad$ $2x - 7 < -5$
$2x - 7 + 7 > 5 + 7$ $\qquad$ $2x - 7 + 7 < -5 + 7$
$2x > 12$ $\qquad$ $2x = 2$
$x > 6$ $\qquad$ $x < 1$

Solution: $x > 6$ or $x < 1$

In interval form: $(-\infty, 1) \cup (6, \infty)$

Graphically:

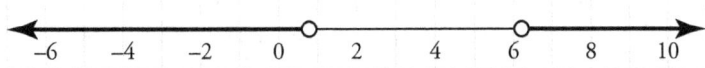

Functions

An equation like $y = 3x + 5$ describes a relation between the independent variable x and the dependent variable y. Thus y is written as $f(x)$, "function of x." But y is only a true **function** if there is a relationship between the set of all inputs or values of the independent variable (the domain) and the set of all outputs or values of the dependent variable (the range) such that each element of the domain corresponds to one element of the range. (For any input there is exactly one output.)

Example:

x	y
2	4
4	8
8	16
(This is a function.)	

x	y
3	7
3	10
6	13
(This is not a function.)	

Example: Given the function $f(x) = 3x + 5$:

Find $f(2)$; $f(0)$; $f(-10)$

Finding $f(2)$ means finding the function value at $x = 2$.

For $f(x) = 3x + 5$:

$$f(2) = 3(2) + 5 = 6 + 5 = 11$$

$$f(0) = 3(0) + 5 = 0 + 5 = 5$$

$$f(-10) = 3(-10) + 5 = -30 + 5 = -25$$

Composition of Functions

Composition of functions is a way of combining functions such that the range of one function is the domain of another. For instance, the composition of functions f and g can be either $f \circ g$ (the composite of f with g) or $g \circ f g + f$ (the composite of g with f). Another way of writing these compositions is $f(g(x))$ and $g(f(x))$. The domain of the composition $f(g(x))$ includes all values x such that $g(x)$ is in the domain of $f(x)$.

Example: Given $f(x) = 3x + 5$ and $g(x) = x^2$, find $f(g(-3))$

Working from the inside out, first evaluate g for -3: $g(-3) = (-3)^2 = 9$

Then insert into the f function: $f(g(-3)) = f(9) = 3(9) + 5 = 32$

Example: What is the composition $f \circ g$ for the functions $f(x) = ax$ and $g(x) = bx^2$?

The correct answer can be found by substituting the function $g(x)$ into $f(x)$: $f(g(x)) = a \cdot g(x) = abx^2$

On the other hand, the composition $g \circ f$ would yield a different answer.

$$g(f(x)) = b \cdot (f(x))^2 = b(ax)^2 = a^2bx^2$$

Inverses of Functions

The **inverse** of a function $f(x)$ is typically labeled $f^{-1}(x)$ and satisfies the following two relations:

$$f(f^{-1}(x)) = x$$
$$f^{-1}(f(x)) = x$$

For a function $f(x)$ to have an inverse, it must be one-to-one. This fact is easily seen, since both $f(x)$ and $f^{-1}(x)$ must satisfy the vertical line test (that is, both must be functions). A function takes each value in a domain and relates it to only one value in the range. Logically, then, the inverse must do the same, only backwards: relate each value in the range to a single value in the domain.

Finding Inverses of Functions

Finding the inverse of a function can be a difficult or impossible task, but there are some simple approaches that can be followed in many cases. The simplest method for finding the inverse of a function is to interchange the variable and the function symbols and then solve to find the inverse. The approach is summarized in the outline below, given a one-to-one function $f(x)$.

1. Replace the symbol $f(x)$ with x
2. Replace all instances of x in the function definition with $f^{-1}(x)$ (or y or some other symbol)
3. Solve for $f^{-1}(x)$.
4. Check the result using $f(f^{-1}(x)) = x$ or $f^{-1}(f(x)) = x$.

Example: Determine if the function $f(x) = x^2$ has an inverse. If so, find the inverse.

First, determine if $f(x)$ is one-to-one. Note that $f(1) = f(-1) = 1$, so $f(x)$ is not one-to-one and therefore has no inverse function.

Example: Determine if the function $f(x) = x^3 + 1$ has an inverse. If so, find the inverse.

The function $f(x) = x^3 + 1$ has an inverse because it increases monotonically for $x > 0$ and decreases monotonically for $x < 0$. As a result, it is one-to-one, and the inverse exists. To calculate the inverse, let y be $f^{-1}(x)$. Replace $f(x)$ with x and replace x with y.

$$f(x) = x^3 + 1 \rightarrow x = y^3 + 1$$

Solve for y. $x - 1 = y^3$

$y = \sqrt[3]{x-1}$

$f^{-1}(x) = \sqrt[3]{x-1}$

Test the result.

$f^{-1}(f(x)) = \sqrt[3]{(x^3+1)-1}$
$= \sqrt[3]{x^3+1-1} = \sqrt[3]{x^3} = x$

The result is thus correct.

Properties of Functions

A **relation** is any set of ordered pairs. The domain of a relation is the set containing all the first coordinates of the ordered pairs, and the range of a relation is the set containing all the second coordinates of the ordered pairs.

A function is a relation in which each value in the domain corresponds to only one value in the range. It is notable, however, that a value in the range may correspond to any number of values in the domain. Thus, although a function is necessarily a relation, not all relations are functions, since a relation is not bound by this rule.

On a graph, use the vertical line test to check whether a relation is a function. If any vertical line intersects the graph of a relation in more than one point, as in the graph below, then the relation is not a function.

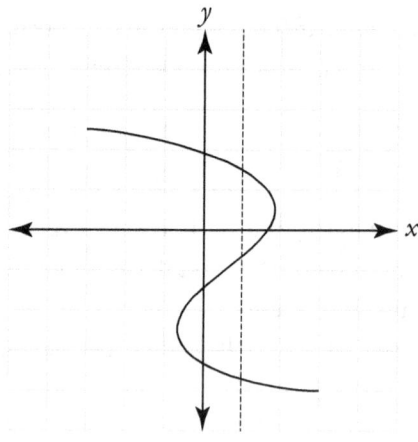

A relation is considered one-to-one if each value in the domain corresponds to only one value in the range, and each value in the range corresponds to only one value in the domain. Thus, a one-to-one relation is also a function, but it adds an additional condition.

In the same way that the graph of a relation can be examined using the vertical line test to determine whether it is a function, the horizontal line test can be used to determine if a function is a one-to-one relation. If no horizontal lines superimposed on the plot intersect the graph of the relation in more than one place, then the relation is one-to-one (assuming it also passes the vertical line test and, therefore, is a function).

As mentioned above, a function is a relation in which each value in the domain corresponds to only one value in the range. Functions can be expressed discretely, as sets of ordered pairs, or they can be expressed more generally as formulas. For instance, the function $y = x$ is a function that represents an infinite set of ordered pairs (x, y), where each value in the domain (x) corresponds to the same value in the range (y).

Families of Functions

Some of the most commonly used function families include linear, polynomial, rational, exponential, logarithmic, and trigonometric functions. These functions, separately or in various combinations, can be used to model a range of common phenomena in finance, physics, and other fields.

Linear Functions

A **linear function** can be expressed as $f(x) = mx + b$, where m and b are constants. It is called linear because it involves no quadratic or cubic variables, nor any square roots or cube roots of variables. No variables in a linear function have any exponent other than 1.

A linear function can be graphed as $y = mx + b$. The result is a straight line with slope m that intercepts the y-axis at $(0, b)$.

Example: $y = 2x - 1$

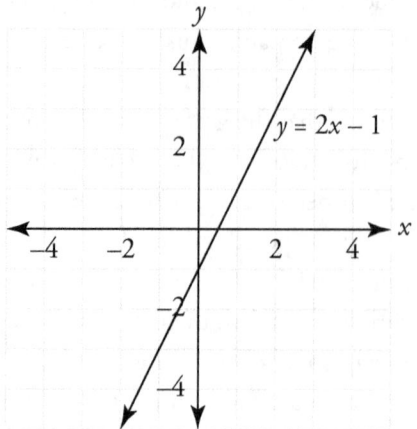

Example: A brand of ketchup contains 3g of sugar per ounce. This is a direct variation. If the ounces of ketchup are plotted as x and the grams of sugar as y, then $y = 3x$ as in the graph below.

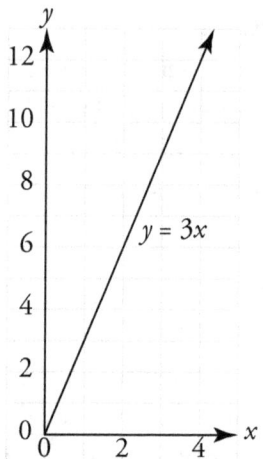

Direct Variation

If, in a function $y = mx + b$, b is 0, the relation is a direct variation. in a direct variation, y and x are always in the same proportion. that means that there is a constant c such that theoretically, c could be positive or negative, but in actual practice, c is almost always positive, which means that as one parameter gets larger, so does the other. The graph of a direct variation always passes through (0, 0).

Domain and Range

The **domain** of a function is the set of all possible inputs to the function. The **range** of a function is the set of all possible outputs. In some functions, both the domain and the range extend to all real numbers. Some functions have limitations on the domain, meaning that certain values are not allowed as inputs. Some functions have limitations on the range, meaning that certain values are not possible as outputs.

In the function $y = 2x + 4$, both the domain and the range extend to all real numbers. Any real number is a possible value of the input x or the output y.

In the function $y = \frac{1}{x-3}$, the domain includes all real numbers except 3: x cannot equal 3, because that would cause a division by zero.

In the function $y = x^2$, the range is all positive real numbers. Since the square of a real number is always positive, the possible outputs of the function do not include any negative numbers.

Intercepts

The **intercepts** of a function are the points at which the function crosses the x- or y-axis. Since the x-value of any point on the y-axis is 0, the y-intercept of any function can be found by setting x equal to 0 and using the function to find the corresponding y-value.

Example: find the y-intercept of $f(n) = x^2 + x + 4$

Let $x = 0$. $f(0) = 0^2 + 0 + 4 = 4$.

The y-intercept is at $(0, 4)$.

Similarly, setting $f(x)$ equal to 0 and solving for x makes it possible to find an x-intercept. Whereas a function normally has one y-intercept, a function can have 0, 1, or multiple x-intercepts.

Example: find any x-intercepts for the function $f(x) = x^2 - 25$

Let $f(x) = 0$. $0 = x^2 - 25$

$x^2 = 25$

$x = \pm 5$

There are x-intercepts at $(5, 0)$ and $(-5, 0)$.

Systems of Linear Equations

A system of equations is a group of equations which are simultaneously true. These equations may involve one, two or more variables. A system of equations that has the same number of equations as there are variables is called a "square system". A square system will have unique solutions for each variable. A nonsquare system may have no solution at all or infinitely many solutions. We will explore three basic ways of solving systems of linear equations: graphing, elimination, and substitution.

Solving a System of Equations by Graphing

A system of equations in two variables can be solved by solving each equation for y and graphing each equation on a common set of axes. The intersection point is the solution for x and y.

Example: solve by graphing.

$$\begin{cases} 2x - y = 1 & [1] \\ x + y = 2 & [2] \end{cases}$$

Solving [1] for y gives $y = 2x - 1$. Solving [2] for y gives $y = -x + 2$. Plot the two lines.

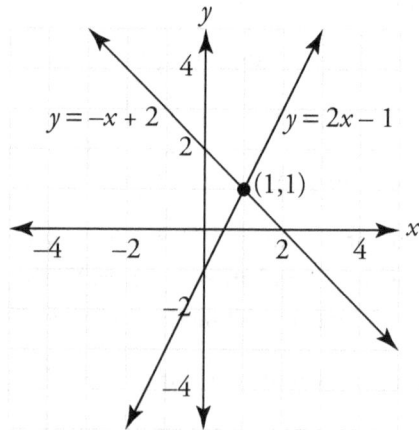

The intersection point appears to be at (1, 1), indicating a solution of $x = 1$, $y = 1$.

Because solving by graphing depends on the accuracy of visual inspection, solving by elimination or substitution is more reliable.

Solving a System of Equations by Elimination

Here we use a method which results in the elimination of one variable; this reduces the system to one equation.

Example: Solve by the elimination method.

$$\begin{cases} 3x + 5y = 10 & [1] \\ 2x + 3y = 7 & [2] \end{cases}$$

Multiply $3x$ by 2 to get $6x$, multiply $2x$ by -3 to get $-6x$.

Add the new equations to eliminate x.

Eq. 1 × 2 $\Rightarrow$ $6x + 10y = 20$ [3]

Eq. 2 × -3 $\Rightarrow$ $-6x - 9y = -21$ [4]

Adding [3] and [4], we get $y = -1$ [5]

Substitute [5] into [1] and solve for x

$$3x + 5(-1) = 10$$
$$3x - 5 = 10$$
$$3x = 15$$
$$x = 5$$

Check solutions using equation [2]:

$2(5) + 3(-1) = 7$

$10 - 3 = 7$ ✓

Solution is $x = 5, y = -1$.

The answer can be represented by $(5, -1)$, since it represents the point of intersection of the two lines.

Solving a System of Equations by Substitution

Here we rewrite one equation in terms of a single variable. Then we substitute the expression of the variable into the second equation.

Example: Solve the system by the substitution method:

$$\begin{cases} x - y = 1 & [1] \\ 3x + 2y = 38 & [2] \end{cases}$$

from [1], adding y to each side produces

$$x = y + 1 \qquad [3]$$

substitute [3] into [2]

$$3(y+1) + 2y = 38$$
$$3y + 3 + 2y = 38$$
$$5y + 3 = 38$$
$$5y = 35$$
$$y = 7 \qquad [4]$$

Substituting [4] into 3] produces

$$x = 7 + 1 = 8$$

The solution is $x = 8$, $y = 7$ or (8,7).

Note: When solving a linear system of equations the elimination method is easier to use for most if not all problems.

Systems of Equations with Infinitely Many Solutions

If a system of equations reduces to a true statement, such as 3 = 3, the system has infinitely many solutions.

Example: Solve the system

$$\begin{cases} 2m + 5n = 1 & [1] \\ 6m + 15n = 3 & [2] \end{cases}$$

Multiply [1] by -3: $-6m - 15n = -3$ \qquad [3]

Adding [3] and [2] gives

0 = 0, which is true for any values of x. Therefore, there are infinitely many solutions. The two equations represent the same line.

Systems of Equations with No Solution

Example: Solve the system

$$\begin{cases} 7x + 5y = 25 & [1] \\ 14x + 10y = -30 & [2] \end{cases}$$

Multiplying [1] by −2 gives

$-14x - 10y = -50$ [3]

Adding [2] and [3] gives

$0 = -80$, which is false.

Therefore, there is no solution. The two lines are parallel and do not intersect.

Systems of Linear Inequalities

Solving systems of linear inequalities is best performed graphically. To graph a linear inequality expressed in terms of x and y, solve the inequality for y. This renders the inequality in slope-intercept form ($y = mx + b$). The point $(0, b)$ is the y-intercept, and m is the slope of the line. If the inequality is expressed only in terms of x, solve for x. When solving an inequality, remember that dividing or multiplying by a negative number will reverse the direction of the inequality sign.

If an inequality yields any of the following results in terms of y, where a is some real number, the solution set of the inequality is bounded by a *horizontal line*: $y < a$, $y \leq a$, $y > a$, $y \geq a$

If the inequality yields any of the following results in terms of x, then the solution set of the inequality is bounded by a *vertical line*: $x < a$, $x \leq a$, $x > a$, $x \geq a$

When graphing the solution of a linear inequality, the boundary is drawn as a dashed line if the inequality sign is $<$ or $>$. This indicates that points on the line do not satisfy the inequality. If the inequality sign is either $\leq$ or $\geq$, then the boundary is drawn as a solid line to indicate that the points on the line satisfy the inequality.

The line drawn as directed above is only the boundary of the solution set for an inequality. The solutions actually include the half plane bounded by the line.

Since, for any line, half of the values in the full plane (for either x or y) are greater than those defined by the line and half are less, the solution of the inequality always amounts to a half plane. Which half plane satisfies the inequality can be found by testing a point on either side of the line. The solution set can be indicated on a graph by shading the appropriate half plane.

For inequalities expressed as a function of x, shade above the line when the inequality sign is $>$ or $\geq$. Shade below the line when the inequality sign is $<$ or $\leq$.

For inequalities expressed as a function of y, shade to the right for $>$ or $\geq$. Shade to the left for $<$ or $\leq$.

The solution to a system of linear inequalities consists of the portion of the graph where the shaded half planes for all the inequalities in the system overlap. For instance, if the graph of one inequality was shaded with red and the graph of another inequality was shaded with blue, then the overlapping area would be shaded purple. The points in the purple area would be the solution set of this system.

Example: Solve by graphing:

$$\begin{cases} x + y \leq 6 \\ x - 2y \leq 6 \end{cases}$$

Solving the inequalities for y, they become

$$\begin{cases} y \leq -x + 6 \text{ (slope } -1, y\text{-intercept } 6) \\ y \geq \frac{1}{2}x - 3 \text{ (slope } \frac{1}{2}, y\text{-intercept } -3) \end{cases}$$

A graph with the appropriate shading is shown below:

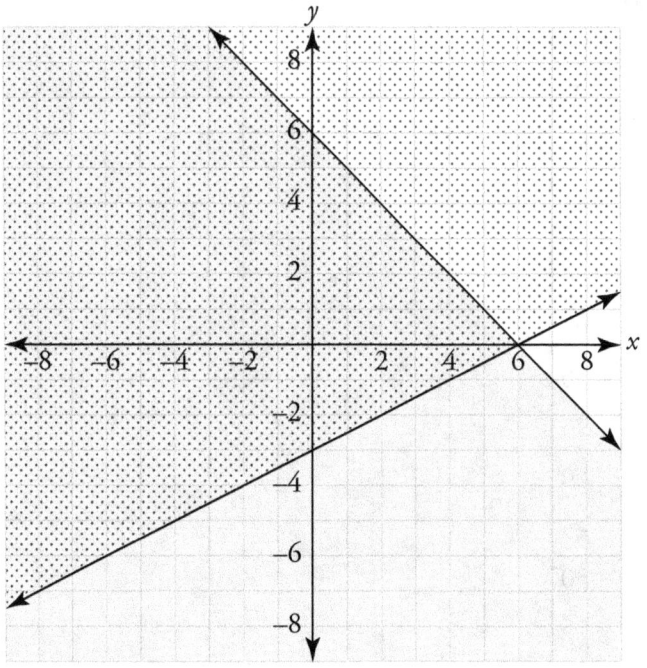

Modeling Functions

Functions can be represented in a variety of ways, namely verbally, symbolically, and graphically. The symbolic expression need simply be evaluated for a representative set of points that can be used to produce a sufficiently detailed graph or table. For example, the equation $y = 9x$ describes the relationship between y, the total number of dollars earned, and x, the number of $9 sunglasses sold. In a relationship of this type, one of the quantities (e.g., total amount earned) is dependent on the other (e.g., number of sunglasses sold). These variables are known as the dependent and independent variables, respectively. A table using this data would appear as:

number of sunglasses sold	1	5	10	15
total dollars earned	9	45	90	135

Demonstrate

Each (x, y) relationship between a pair of values is called a coordinate pair and can be plotted on a graph. The coordinate pairs (1, 9), (5, 45), (10, 90), and (15, 135) are plotted on the graph below.

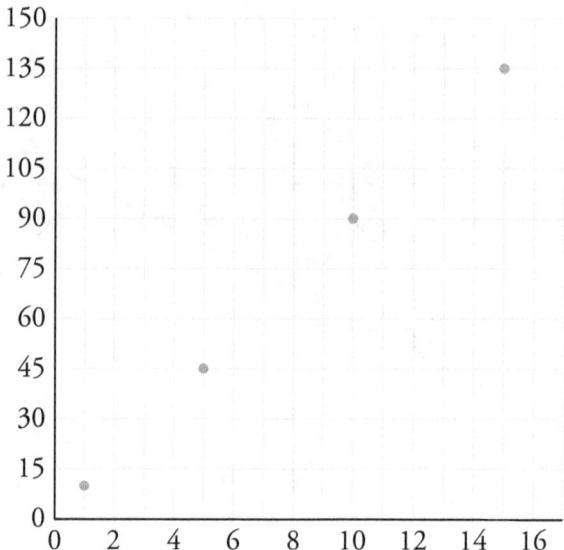

The graph shows a linear relationship. A **linear relationship** is one in which the change in two quantities is in a constant proportion. Doubling the change in x also doubles the change in y. On a graph, a straight line depicts a linear relationship. When the slope of the line is positive, the dependent data is increasing as the independent data increases and the function can be said to represent **linear growth**. Additionally, patterns of growth are usually presented graphically in the first quadrant even though they may additionally exist elsewhere.

Models of growth exist in many other mathematical patterns. Two more notable growth patterns are as follows:

Quadratic Growth $h(x) = ax^2$

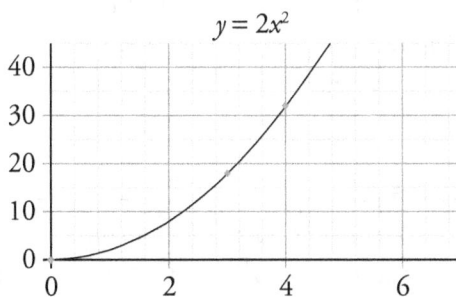

Exponential Growth (with $n > 0$) $f(x) = an^x$

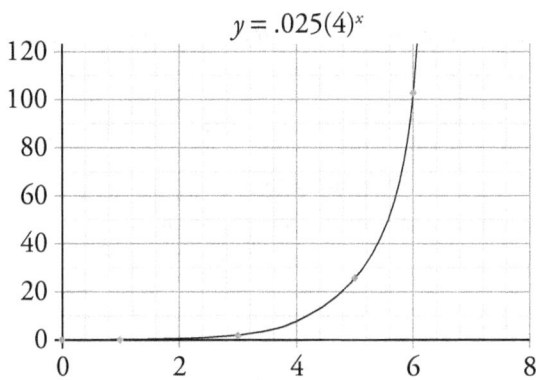

Deriving a (Symbolic) Function from Data

Example: What is the equation that expresses the relationship between x and y in the table below?

x	0	1	2	3	4	5
y	3	5	7	9	11	13

Each x-input differs from the next by a constant, 1. Since each y-output also differs from the next by a constant, 2, these data can be modeled by a linear function. Additionally, a graph of the data would reveal a straight line.

We will write the equation as $y = mx + b$. We can determine m (the slope) as the change in y between any two points divided by the change in x between any two points. In this table, the change in y between two points is 2 and the change in x between two points is 1, so the slope m is $\frac{2}{1} = 2$. Find b by setting x to zero. In this table, when $x = 0$, $y = 3$, so $b = 3$, and the equation is $y = 2x + 3$.

Transformations

Transformations represent the manipulation of objects through movement, rotation, and scaling. The transformed version of an object is called its **image**. If the original object is labeled with letters, such as ABCD, the image can be labeled with the same letters followed by a prime symbol: A′B′C′D′.

Transformations can be characterized in different ways.

Translations

A **translation** is a transformation that "slides" an object a fixed distance in a given direction. The original object and its translation have the same shape and size, and they face in the same direction.

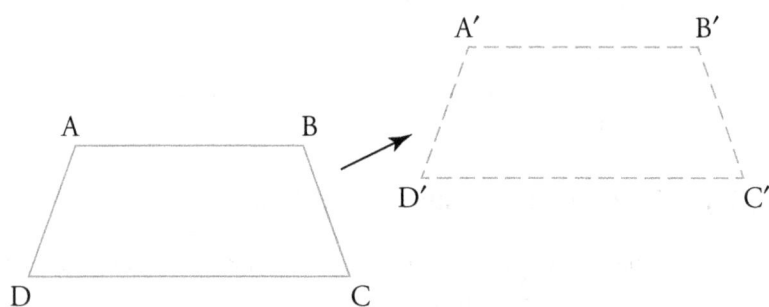

In the coordinate plane, a point or figure can be translated
- h units to the right by adding h to the x-coordinates of all points
- h units to the left by subtracting h from the x-coordinates of all points
- k units up by adding k to the y-coordinates of all points
- k units down by subtracting k from the y-coordinates of all points

Example: A triangle with coordinates $(-1, 2)$, $(3, 5)$ and $(1, -3)$ is translated 5 units left and 1 unit up. What are the coordinates of the translated triangle?

To translate 5 units left, 5 must be subtracted from each x-coordinate. To translate 1 unit up, 1 must be added to each y-coordinate. The coordinates of the new triangle are (–6, 3), (–2, 6) and (–4, –2).

Translating a Function or Equation

The graph of a parabola produced by an equation in the form $y = f(x)$ can be translated
- h units to the right by substituting $(x - h)$ for x
- h units to the left by substituting $(x + h)$ for x
- k units up by substituting $(y - k)$ for y
- k units down by substituting $(y + k)$ for y

Example: Describe the translation from $y = x$ to $y = x + 7$

The line $y = x$ has a slope of one and goes through the origin. The translated line, $y = x + 7$, can be rewritten as $(y - 7) = x$ which predicts a vertical move 7 spaces up, to have a new y intercept at $(0, 7)$. The line will still have a slope of 1.

Example: Graph the function $y = x^2$. Then move it 3 units right, left, up and down.

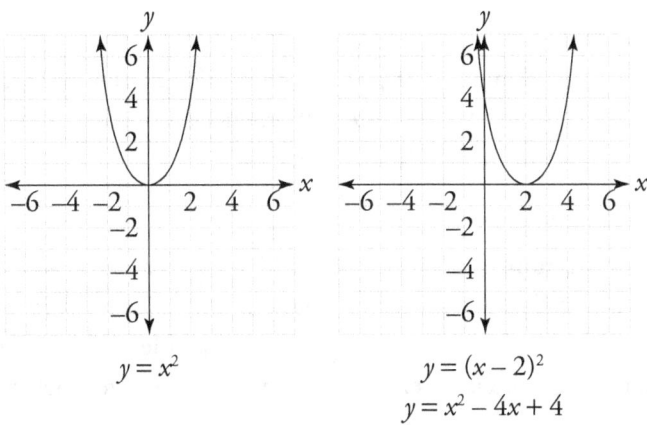

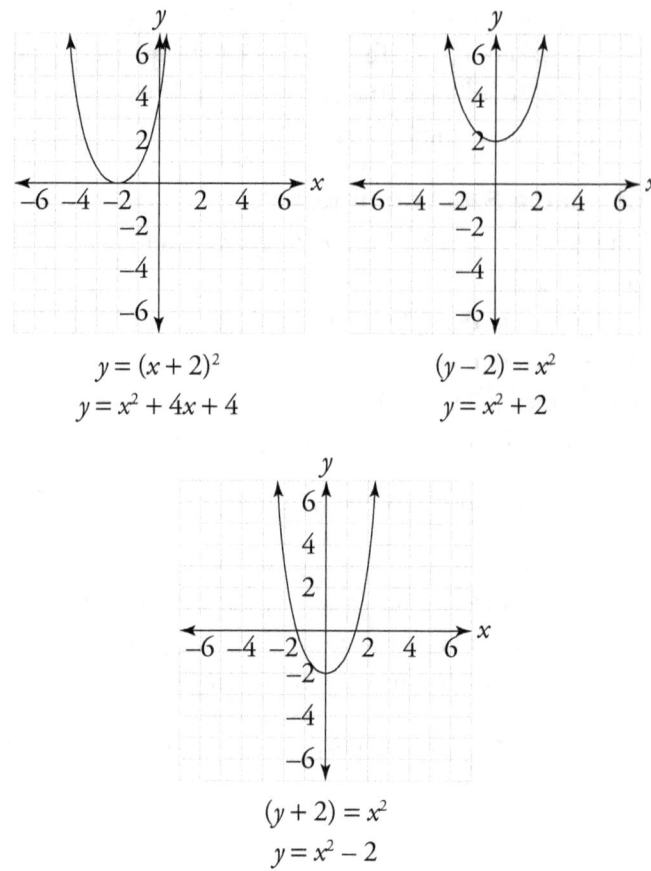

Summary: If $f(x)$ is a quadratic function producing a parabola, then $f(x - h) + k$ will produce a similar parabola shifted h units to the right and k units up.

Rotations

A **rotation** is a transformation that turns a figure about a fixed point, which is called the center of rotation. An object and its rotation are the same shape and size, but the figures may be oriented in different directions. Rotations can occur in either a clockwise or a counterclockwise direction.

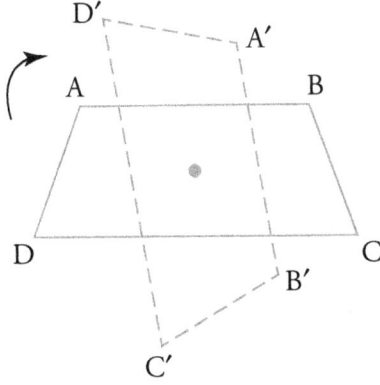

In the coordinate plane, a point or figure can be rotated about the origin

- 90° counterclockwise (or 270° clockwise) by changing (x, y) to $(-y, x)$

    [**Example:** the point $(-3, 2)$, rotated 90° counterclockwise around the origin, produces the point $(-2, -3)$.]

- 180° clockwise or counterclockwise by changing (x, y) to $(-x, -y)$

    [**Example:** the point $(6, -7)$, rotated 180° around the origin, produces the point $(-6, 7)$.]

- 90° clockwise (or 270° counterclockwise) by changing (x, y) to $(y, -x)$

    [**Example:** the point $(-1, 5)$, rotated 90° clockwise around the origin, produces the point $(5, 1)$.]

Example: A triangle with coordinates $(-3, 2), (4, -1), (1, -2)$ is rotated 90° clockwise. What are the coordinates of the rotated triangle?

To produce a 90° clockwise rotation, each x-coordinate must be replaced with its opposite and the two coordinates must be switched.

The coordinates of the new triangle are $(2, 3), (-1, -4), (-2, -1)$.

Reflections

The **reflection** of a figure across a line, called the line of reflection, produces a figure similar but reversed, at an equal distance on the opposite side of the line, as if it were a mirror image and the line of reflection was the mirror.

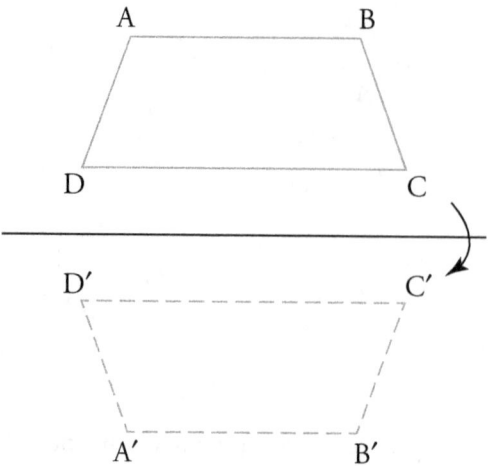

In the coordinate plane, a point or figure can be reflected

- across the y-axis by changing (x, y) to $(-x, y)$

    [**Example:** the point $(-2, 5)$, reflected across the y-axis, becomes $(2, 5)$.]

- across the x-axis by changing (x, y) to $(x, -y)$

    [**Example:** the point $(-4, 5)$, reflected across the x-axis, becomes $(-4, -5)$.]

- across the line $y = x$ by changing (x, y) to (y, x)

    [**Example:** the point $(5, 7)$, reflected across the line $y = x$, becomes $(7, 5)$.]

- across the line $y = -x$ by changing (x, y) to $(-y, -x)$

    [**Example:** the point $(-3, -4)$, reflected across the line $y = -x$, becomes $(4, 3)$.]

Example: A rectangle with coordinates $(0, 3)$, $(0, -2)$, $(4, 3)$, $(4, -2)$ is reflected across the x-axis. What are the coordinates of the reflected rectangle?

To produce a reflection across the x-axis, each y-coordinate must be replaced by its opposite. The new rectangle has coordinates $(0, -3)$, $(0, 2)$, $(4, -3)$, $(4, 2)$.

The examples of a translation, a rotation, and a reflection given above are for polygons, but the same principles apply to the simpler geometric elements of points and lines. In fact, a transformation performed on a polygon can be viewed equivalently as the same transformation performed on the set of points (vertices) and lines (sides) that compose the polygon. Thus, to perform complicated transformations on a figure, it is helpful to perform the transformations on all the points (or vertices) of the figure, then reconnect the points with lines as appropriate.

Example: Reflect the line $y = -2x - 5$ over the x-axis.

A table of values for the original line can be as follows:

x	-4	-2.5	0	1
y	3	0	-5	-7

Since a set of points reflected over the x axis replaces every y with its opposite, the table of values for the reflection converts to:

x	-4	-2.5	0	1
y	-3	0	5	7

When graphed, these sets of data create two lines that appear symmetrical with respect to the x axis.

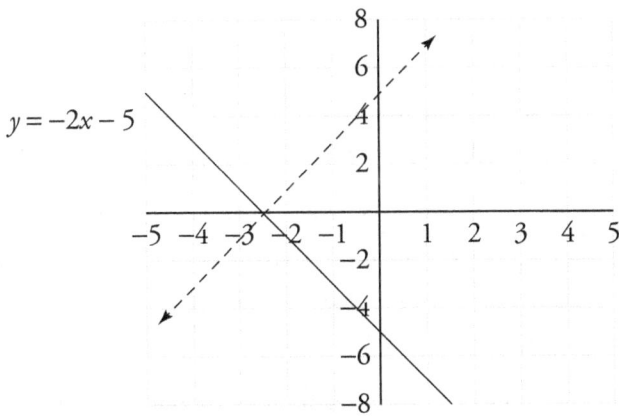

To find the equation of the reflection, take the original equation and replace y with (–y). Then solve for the "new" y.

$$y = -2x - 5 \rightarrow -y = -2x - 5$$

$y = 2x + 5$ is the equation of the reflection.

Multiple Transformations

Multiple transformations can be performed on a geometrical figure. The order of these transformations may or may not be important. For instance, multiple translations can be performed in any order, as can multiple rotations (around a single fixed point) or reflections (across a single fixed line). The order of the transformations becomes important when several types of transformations are performed or when the point of rotation or the line of reflection changes among transformations. For example, consider a translation of a given distance upward and a clockwise rotation by 90° around a fixed point. Changing the order of these transformations changes the result.

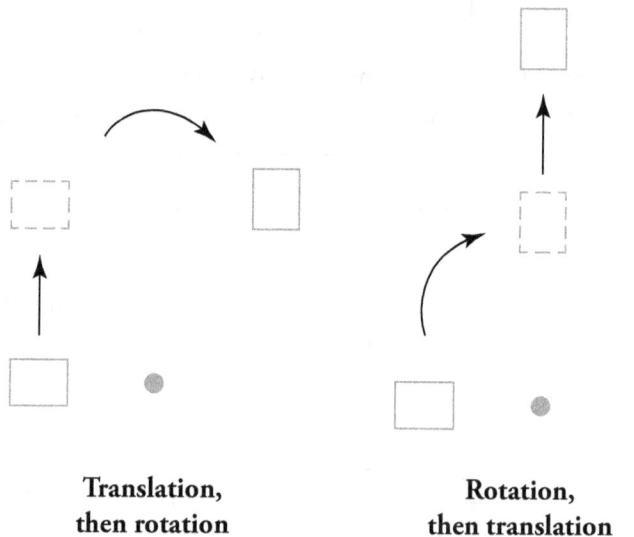

**Translation,
then rotation**

**Rotation,
then translation**

As shown, the final position of the box is different, depending on the order of the transformations. Thus, it is crucial that the proper order of transformations (whether determined by the details of the problem or some other consideration) be followed.

Example: Find the final location of a point at (1, 1) that undergoes the following transformations: rotate 90° counterclockwise about the origin; translate distance 2 in the negative y-direction; reflect about the y-axis.

First, draw a graph of the x- and y-axes and plot the point at (1, 1).

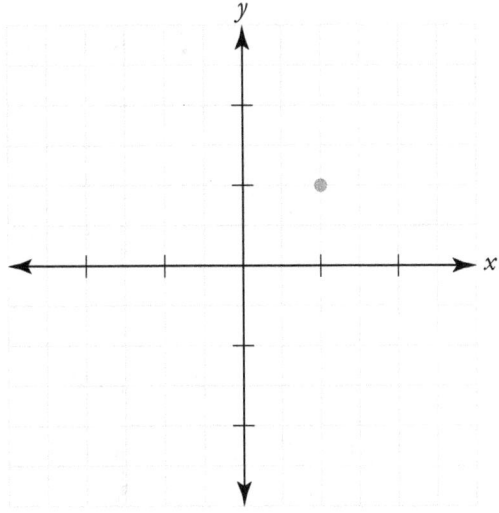

Next, perform the rotation. The center of rotation is the origin, and the rotation is in the counterclockwise direction. In this case, the even value of 90° makes the rotation simple to do by inspection. Next, perform a translation of distance 2 in the negative y direction (down). The results of these transformations are shown below.

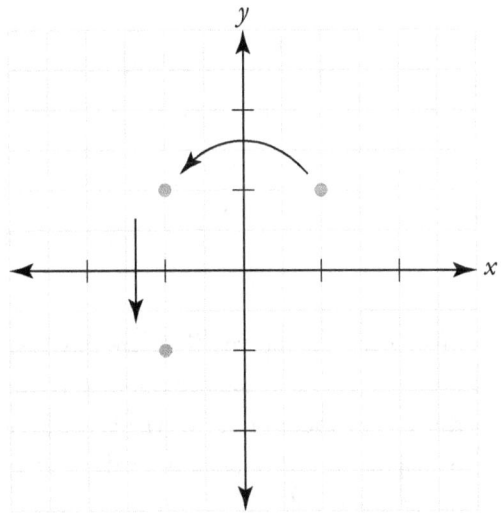

Finally, perform the reflection about the y-axis. The final result, shown below, is a point at (1, −1).

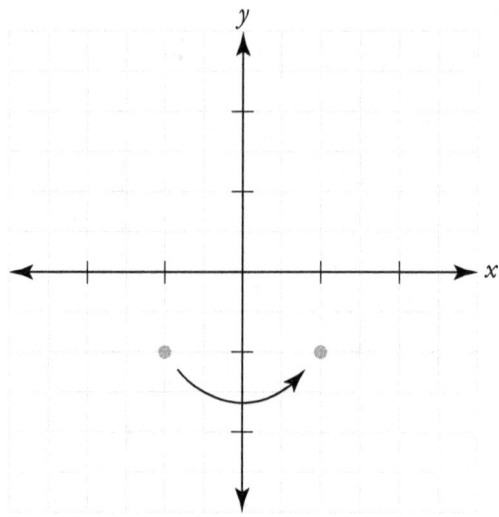

Using this approach, polygons can be transformed on a point-by-point basis. For some problems, there is no need to work with coordinate axes. For instance, the problem may simply require transformations without respect to any absolute positioning.

Example: Rotate the following regular pentagon by 36° about its center, and then reflect it about a horizontal line.

First, perform the rotation. In this case, the direction is not important because the pentagon is symmetric. As it turns out in this case, a rotation of 36° yields the same result as flipping the pentagon vertically (assuming the vertices of the pentagon are indistinguishable).

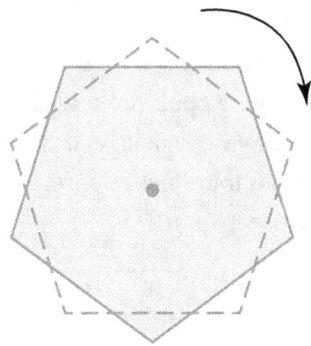

Finally, perform the reflection. Note that the result here is the same as a downward translation (assuming the vertices of the pentagon are indistinguishable).

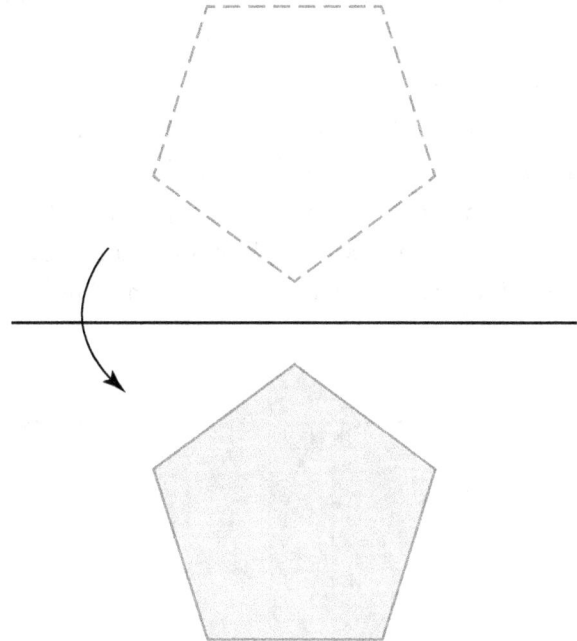

Symmetry

Symmetry is a property of a figure whose parts can be made to match up perfectly in a changed position. There are two kinds: reflectional symmetry and rotational symmetry.

Reflectional Symmetry

A figure has **reflectional symmetry** if there is a line, called a **line of symmetry**, that the figure could be folded across so that the halves would match completely. A figure can have more than one line of symmetry. A square, for instance, has four lines of symmetry, as the figure shows.

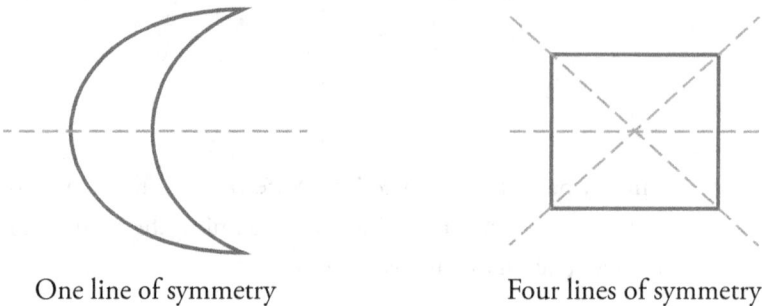

One line of symmetry   Four lines of symmetry

Rotational Symmetry

A figure has **rotational symmetry** if it would look the same after being rotated less than 360°. If the figure is unchanged by a rotation of 180°, its rotational symmetry can be described either as **order 2 rotational symmetry** (because the figure can appear in 2 identical positions), or as **180° rotational symmetry**. The recycling symbol below has rotational symmetry of order 3 or 120°.

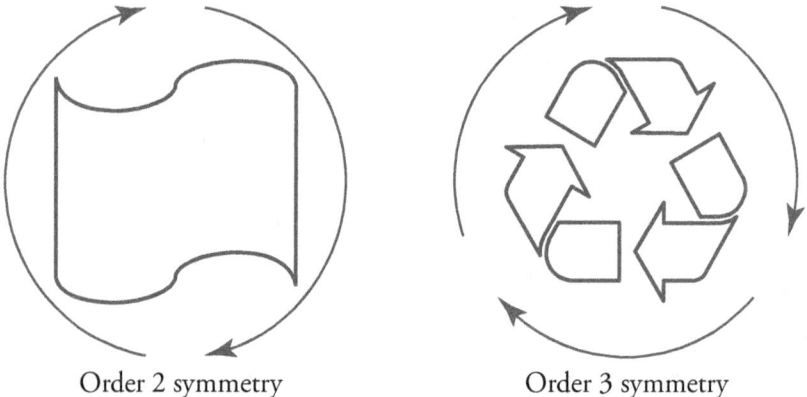

Order 2 symmetry   Order 3 symmetry

Symmetry on the Coordinate Plane

Symmetries in a function can also be described in terms of reflections or "mirror images." A function can be symmetric about the y-axis (but not about the x-axis, except for the function $f(x) = 0$, since every function must pass the vertical line test). A function is symmetric about the y-axis if for every point (x, y) that is included on the graph of the function, the point $(-x, y)$ is also included on the graph. Consider the function $f(x) = x^2$. Note that for each point (x, x^2) on the graph, the point $(-x, x^2)$ is also on the graph. The symmetry of the function about the y-axis can also be seen in the graph below.

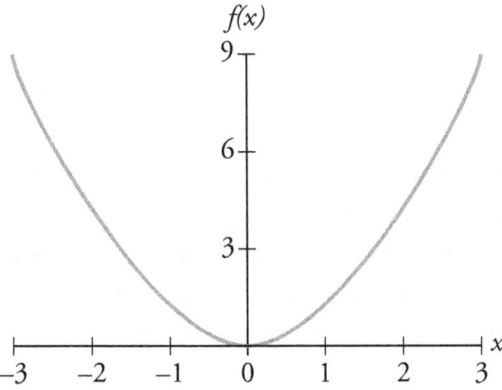

A function that is symmetric about the y-axis is also called an even function. Although functions cannot be symmetric about the x-axis, relations that do not obey the vertical line test can be symmetric in this way. A relation is symmetric about the x-axis if for every point (x, y) in the graph of the relation, the point $(x, -y)$ is also in the graph.

Consider, for instance, the relation Ï $g(x) = \pm\sqrt{x}$. For every value of x in the domain, the points $(x, \sqrt{x})$ and $(x, -\sqrt{x})$ are both in the graph, as shown below.

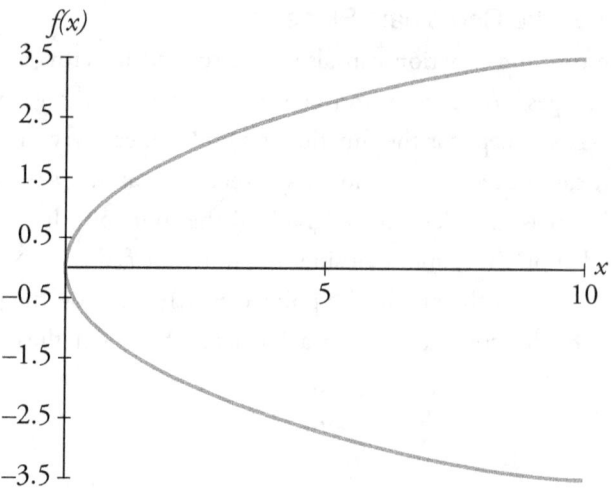

Functions may also be rotationally symmetric with respect to the origin. Such functions are called **odd (or antisymmetric) functions** and are defined by the property that for any point (x, y) on the graph of the function, the point $(-x, -y)$ is also on the graph of the function. The function $f(x) = x^3$, for instance, is rotationally symmetric with respect to the origin, as shown in the graph below.

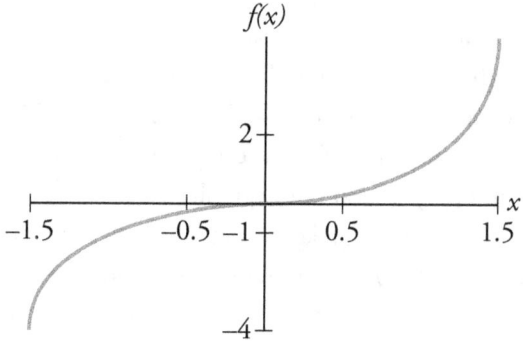

Counting and Probability

Sample Space Size

In probability, the **sample space** is a list of all possible outcomes of an experiment. For example, the sample space of tossing two coins is the set {HH, HT, TT, TH} where heads is H and tails is T and the sample space of rolling a six-sided die is the set {1, 2, 3, 4, 5, 6}.

When analyzing experiments with a large number of possible outcomes, it is important to find the size of the sample space. The size of the sample space can be determined by using the fundamental counting principle and the rules of combinations and permutations.

- The **fundamental counting principle** states that if there are m possible outcomes for one task and n possible outcomes of another, there are $(m \times n)$ possible outcomes of the two tasks together.
- A **permutation** is one of the possible arrangements of n items, without repetition, where the order of selection is important.
- A **combination** is one of the possible arrangements of n items, without repetition, where the order of selection is not important.

Example: How many different ice cream sundaes can be created when selecting from 3 different flavors of ice cream and 4 different sauce toppings either with or without nuts?

By the fundamental counting principle, the number of possible outcomes is $3 \times 4 \times 2 = 24$.

Example: If any two numbers are selected from the set {1, 2, 3, 4}, list the possible permutations and combinations.

Combinations	Permutations
12, 13, 14, 23, 24, 34	12, 21, 13, 31, 14, 41, 23, 32, 24, 42, 34, 43
six ways	twelve ways

Note: that the list of permutations includes 12 and 21 as separate possibilities, since the order of selection is important. In the case of combinations, however, the order of selection is not important and, therefore, 12 is the same combination as 21. Hence, 21 is not listed separately as a possibility.

The number of permutations and combinations may also be found by using the formulae given below.

- The number of possible permutations in selecting *r* objects from a set of *n* objects is given by $_nP_r = \dfrac{n!}{(n-r)!}$.

The notation $_nP_r$ is read "the number of permutations of *n* objects taken *r* at a time."

Note: that **n!** represents the **factorial** of a positive integer, n, such that n! is the product of that integer and every lesser positive integer down to 1. The factorial of 5, for instance, which is written as 5!, equals $5 \times 4 \times 3 \times 2 \times 1$. By convention, the factorial of 0 is 1.

In our example, two objects are being selected from a set of four.

$$_4P_2 = \dfrac{4!}{(4-2)!} = \dfrac{4 \times 3 \times 2 \times 1}{2 \times 1} = \dfrac{24}{2} = 12$$

- The number of possible combinations in selecting *r* objects from a set of *n* objects is given by $_nC_r = \dfrac{n!}{(n-r)!r!}$

In our example, $_4C_2 = \dfrac{4!}{(4-2)!2!} = \dfrac{24}{2(2)} = 6$

The fundamental counting principal combines with the concept of permutations to determine how many different ways a set of objects can be arranged in a row. It can be shown that $_nP_n$, the number of ways *n* objects can be arranged in a row, is equal to *n*!. We can imagine *n* positions being filled, one at a time. The first position can be filled in *n* ways using any one of the *n* objects.

Since one of the objects has been used, the second position can be filled in only (*n* – 1) ways. Similarly, the third position can be filled in (*n* – 2) ways, and so on. Hence, the total number of possible arrangements of *n* objects in a row is given by $_nP_n = n(n-1)(n-2)\ldots 1 = n!$

Example: Five books are placed in a row on a bookshelf. In how many different ways can they be arranged?

The number of possible ways in which 5 books can be arranged in a row is $5! = 5 \times 4 \times 3 \times 2 \times 1 = 120$.

The formula given above for $_nP_r$, the number of possible permutations of r objects selected from n objects, can also be proved in a similar manner. If r positions are filled by selecting from n objects, the first position can be filled in n ways, the second position can be filled in $n - 1$ ways, and so on (as shown before). The rth position can be filled in $n - (r - 1) = n - r + 1$ ways. Hence,
$$_nP_r = n(n-1)(n-2)\ldots(n-r+1) = \frac{n!}{(n-r)!}$$

The formula for the number of possible combinations of r objects selected from n objects, $_nC_r$, may be derived by using the above two formulae. For the same set of r objects, the number of permutations is $r!$. All these permutations, however, correspond to the same combination. Hence,
$$_nC_r = \frac{_nP_r}{r!} = \frac{n!}{(n-r)!\,r!}$$

Probability Calculations

The **probability** of an outcome, given a random experiment (a structured, repeatable experiment for which the outcome cannot be predicted or, alternatively, for which the outcome is dependent on "chance"), is the relative frequency of the outcome. The relative frequency of an outcome is the fraction or percentage of times an experiment yields that outcome for a very large (ideally, infinite) number of trials. For instance, if a fair coin is tossed a very large number of times, then the relative frequency of tossing heads is 0.5, or 50% (that is, one out of every two tosses, on average, should come out heads up). The probability is this relative frequency.

Example: What is the probability that an even number will be rolled with the toss of a die?

Given a standard, 6-sided die, 3 sides, namely 2, 4, and 6, fulfill the desired outcome out of the six sides of the die. Therefore, $P(\text{even}) = \frac{3}{6} = \frac{1}{2}$

Example: Given a jar containing 5 chocolate candies, 4 butterscotch, and 2 peppermint, what is the probability of selecting a peppermint? And of NOT selecting a peppermint?

A total of 11 candies fill the jar with 2 of them peppermint. Therefore $P(p) = \frac{2}{11}$. When considering $P(\text{not } p)$, the other 9 candies fulfill this outcome, resulting in a probability of $\frac{9}{11}$. The outcomes of peppermint and "not peppermint" are opposites of each other and are called **complementary events**.

Note: $P(p) + P(\text{not } p) = \frac{2}{11} + \frac{9}{11} = 1$. The probabilities of complementary events always add to 1.

A **random variable** is a function that corresponds to the outcome of some experiment or event, which is in turn dependent on "chance." For instance, the result of a tossed coin is a random variable: the outcome is either heads or tails, and each outcome has an associated probability. A **discrete variable** is one that can only take on certain specific values. For instance, the number of students in a class can only be a whole number (e.g., 15 or 16, but not 15.5). A **continuous variable**, such as the weight of an object, can take on a continuous range of values.

The probabilities for the possible values of a random variable constitute the probability distribution for that random variable. Probability distributions can be discrete, as with the case of the tossing of a coin (there are only two possible distinct outcomes), or they can be continuous, as with the outside temperature at a given time of day. In the latter case, the probability is represented as a continuous function over a range of possible temperatures, and finite probabilities can only be measured in terms of ranges of temperatures rather than specific temperatures. That is to say, for a continuous distribution, it is not meaningful to say "the probability that the outcome is x"; instead, only "the probability that the outcome is between x and $x + k$" is meaningful. (Note that if each potential outcome in a continuous distribution has a nonzero probability, then the sum of all the probabilities would be greater than 1, since there are an infinite number of potential outcomes.)

Example: Find the sample space and construct a probability distribution for the sum of the outcomes from tossing two six-sided dice (each with numbers 1 through 6).

The sample space is the set of all possible outcomes that can arise in a given trial. For two dice, the outcome of any toss can be written as a two-digit number, with each digit being from 1 to 6. Using this convention, the sample space for tossing two 6-sided dice is as follows:

{11, 12, 13, 14, 15, 16,
 21, 22, 23, 24, 25, 26,
 31, 32, 33, 34, 35, 36,
 41, 42, 43, 44, 45, 46,
 51, 52, 53, 54, 55, 56,
 61, 62, 63, 64, 65, 66}

Notice that all the numbers in any diagonal rising from left to right have the same sum. We can show the frequency of the different sums in a histogram:

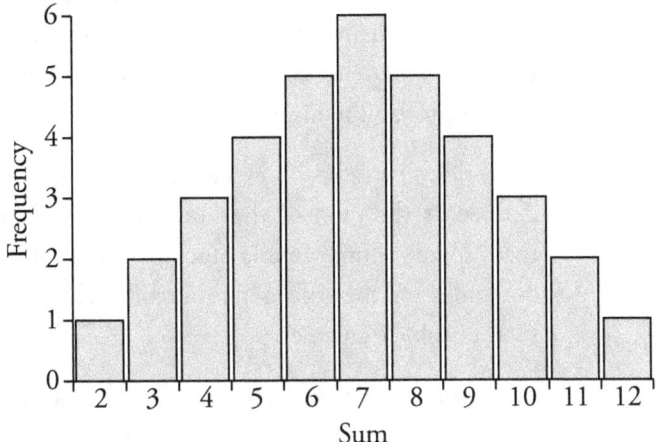

To construct the associated probability distribution, note first that the sum of the probabilities must equal 1. Since there are 36 possibilities in all, the probability of each sum occurring is its frequency divided by 36. Since the sum 6 occurs 5 times in the frequency distribution, the probability of rolling two dice that add to 6 is 5 out of 36 or $\frac{5}{36}$ or about 0.14.

The probability distribution can be shown as a histogram below.

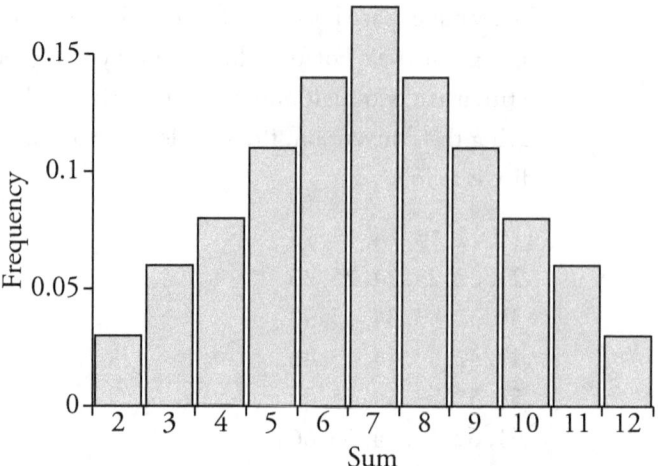

The sum of the probabilities for all the possible outcomes of a discrete distribution must be equal to unity. The **expected value** of a probability distribution is the same as the mean value of a probability distribution. The expected value is thus a measure of the central tendency or average value for a random variable with a given probability distribution.

Probability can also be expressed in terms of odds. **Odds** are defined as the ratio of the number of favorable outcomes to the number of unfavorable outcomes. The sum of the favorable outcomes and the unfavorable outcomes should always equal the total possible outcomes.

For example, given a bag of 12 red marbles and 7 green marbles, compute the odds of randomly selecting a red marble.

Odds of getting red are 12 : 7.
Odds of not getting red are 7 : 12.

In the case of flipping a coin, it is equally likely that a head or a tail will be tossed, so the odds of tossing a head are 1 : 1. This is called even odds.

Conditional probability is the probability that a second event will happen, given that a first event has happened.

Example: Consider the following two events: the home team wins the semifinal round (event A) and the home team wins the final round (event B). The probability of event B is contingent on the probability of event A. If the home team fails to win the semifinal round, it has a zero probability of winning in the final round. On the other hand, if the home team wins the semifinal round, then it may have a finite probability of winning in the final round. Symbolically, the probability of event B given event A is written $P(B|A)$.

Conditional probability can be calculated according to the following definition (the symbol ∩ means "and," the symbol ∪ < means "or," and $P(x)$ means "the probability of x"):

$$P(B|A) = \frac{P(A \cap B)}{P(A)}$$

Example: If the first roll of a die with numbers from 1 to 6 is a 3, what is the probability that the total of two rolls will be 6 or greater?

The probability of event A, rolling a 3, is 1 in 6 or $\frac{1}{6}$. For event A and B both to happen, the outcome of A must be a 3 and the outcome of B must be 3, 4, 5, or 6, four possibilities out of 6 or $\frac{4}{6}$. The probability of both events happening, therefore is

$$\frac{1}{6}\left(\frac{4}{6}\right) = \frac{4}{36} = \frac{1}{9}$$

The probability of B happening in the event that A happens, then, is

$$P(B|A) = \frac{P(A \cap B)}{P(A)} = \frac{1}{9}\frac{1}{6} = \frac{1}{9} \times \frac{6}{1} = \frac{6}{9} = \frac{2}{3}$$

Two events are **independent** if the probability of the second event does not depend on the outcome of the first event. The probability that two independent events will both happen can be found by multiplying the separate probabilities: $P(A \cap B) = P(A)P(B)$

Example: Consider a pair of dice: one red and one green. First the red die is rolled, followed by the green die. What is the probability of rolling a 2 with the red die and a 5 with the green die?

It is apparent that these events do not depend on each other, since the outcome of the roll of the green die is not affected by the outcome of the roll of the red die. Thus the events are independent events.

$$P(A \cap B) = P(A)P(B) = \frac{1}{6}\left(\frac{1}{6}\right) = \frac{1}{36}$$

Often events are not independent. Suppose a jar contains 12 red marbles and 8 blue marbles. If a marble is selected at random and then replaced, the probability of picking a certain color is the same in the second trial as it is in the first trial. If the marble is not replaced, then the probability of picking a certain color is not the same in the second trial, because the total number of marbles is decreased by 1. In this case, the second event is **dependent** on the first, because the first event influences the possibilities of the second event. This is an illustration of conditional probability. If R_1 is the probability of picking a red marble the first time and R_2 is the probability of picking a red marble the second time, then, *if the first marble is replaced before picking the second marble*, the probability of picking a red marble both times is simply the product of the separate probabilities for each of the two trials:

$$P(R_1 \cap R_2) = P(R_1)P(R_2) = \frac{12}{20}\left(\frac{12}{20}\right) = \frac{144}{400} = 0.36$$

If the first marble is *not replaced* before picking the second marble, and a red marble is chosen the first time, in the second trial there will be only 11 red marbles and 19 marbles to choose from. Therefore, the probability of picking a red marble both times equals the probability of picking a red marble in the first trial times the probability of picking a red marble in the second trial, assuming that a red marble has been picked in the first trial.

$$P(R_1 \cap R_2) = P(R_1)P(R_2 \mid R_1) = \frac{12}{20}\left(\frac{11}{19}\right) = \frac{132}{380} \approx 0.347$$

Example: A car has a 75% probability of traveling 20,000 miles without breaking down. It has a 50% probability of traveling 10,000 additional miles without breaking down if it first makes it to 20,000 miles without breaking down. What is the probability that the car reaches 30,000 miles without breaking down?

Let event A be that the car reaches 20,000 miles without breaking down. $P(A) = 0.75$

Event B is that the car travels an additional 10,000 miles without breaking down (assuming it didn't break down for the first 20,000 miles). Since event B is contingent on event A, write the probability as follows: $P(B \mid A) = 0.50$

Use the probability formula for dependent events to find the probability of $A \cap B$, the probability that the car will travel 30,000 miles without breaking down. $P(A \cap B) = P(A)P(B \mid A) = 0.75(0.5) = 0.375$

The car has a 37.5% probability of traveling 30,000 miles without breaking down.

The **expected value** $E(X)$, given a random variable X and an associated probability distribution $f(x)$ is the following for continuous and discrete distributions, respectively:

$$E(X) = \int_{-\infty}^{\infty} x f(x) \, dx$$

$$E(X) = \sum_{i} x_i f(x_i)$$

The expected (or expectation) value for a random variable X is also sometimes written as $\langle X \rangle$. The expected value can be applied to random variables such as the mean μ (written as $E(X)$ or $\langle X \rangle$), the variance (written as $E((X - \mu)^2)$ or $\langle (X - \mu)^2 \rangle$), or any other parameter.

The reasoning behind these formulas comes from the notion that the expected value of a random variable is also the *mean value* of the probability distribution. In the case of discrete variables such as coin toss outcomes, one can think of it as a weighted average. For instance, for a regular six-sided die, where each outcome is equally probable, the expected value on a roll is a simple average and is given by

$$\frac{1+2+3+4+5+6}{6} = \frac{21}{6} = 3.5$$

For a non-standard die, for instance one which even numbers are twice as likely to come up on a roll as odd numbers, the probability of each odd number coming up is 1/9, and the probability of each even number coming up is 2/9. Hence, the expected value of one roll of this die is given by

$$(1+3+5)\left(\frac{1}{9}\right) + (2+4+6)\left(\frac{2}{9}\right) = \frac{33}{9} = 3.67$$

The higher probability of the numbers with relatively greater value is reflected in the increase in the expected value compared to the regular die.

Example: A fair coin is tossed three times. What is the expected value of the total number of heads? Consider the different ways in which a coin can be tossed three times:

HHH, HHT, HTH, THH, TTH, HTT, THT, TTT

Notice that the probability of getting zero heads is 1/8, one head is 3/8, two heads is 3/8, and three heads is 1/8. Hence the expected value for the number of heads is

$$\left(\frac{1}{8}\right)\cdot 0 + \left(\frac{3}{8}\right)\cdot 1 + \left(\frac{3}{8}\right)\cdot 2 + \left(\frac{1}{8}\right)\cdot 3 = \frac{12}{8} = 1.5$$

Data Analysis and Statistics

Measuring Characteristics of Data

It is often desirable to offer some very brief mathematical summary of what a set of data is like. Various ways of summarizing a dataset are available, depending on what quality of the data you wish to show. **Measures of central tendency** show in different ways where the center or the heart of a dataset lies or what sort of data item would be typical. The mean, median, and mode are measures of central tendency (i.e., the average or typical value) in a data set. They can be defined both for discrete and continuous data sets. **Measures of dispersion** show how far the data diverge from the center. Range, variance, and standard deviation are measures of dispersion.

Mean

For discrete data, the **mean** is the average of the data items, or the value obtained by adding all the data values and dividing by the total number of data items.

Example: Find the mean of the following temperatures in degrees Fahrenheit: 71, 82, 65, 93, 87, 79, 82

There are 7 items, so the mean is

$$\frac{71+82+65+93+88+79+82}{7} = \frac{560}{7} = 80$$

Weighted Averages

Sometimes it is desirable to assign more importance to some items in a dataset than to others. That can be done by finding a **weighted average**. Any item in the dataset can be multiplied by a weighting factor k as it is added in to the numerator. That item is also counted, not just as 1 item, but as k items, when tallying the denominator.

Example: Your average in a class is determined by your scores on three quizzes and a final exam. The final exam counts double. Your scores on the quizzes and the final exam, are 83, 91, 86, and 90. What is your average?

Since the final exam counts double, that score is doubled when adding up the total of the scores. That test also counts as 2 tests in the denominator, so the total is divided by 5 instead of 4.

$$\text{Your average is } \frac{83+91+86+2(90)}{1+1+1+2} = \frac{440}{5} = 88$$

Median

The **median** is found by putting the data in order from smallest to largest and selecting the value in the middle (or the average of the two values in the middle if the number of data items is even).

Example: The ages of the members of the school board are 66, 42, 54, 49, 67, and 58. What is the median age? From smallest to largest, the items are 42, 49, 54, 58, 66, and 67. Since there are 6 items, an even number, the median is halfway between the two middle items, 54 and 58, so the median is 56.

Mode

The **mode** is the most frequently occurring data value. There can be more than one mode in a data set, if there is a tie between two or more values for frequency of occurrence.

Example: What is the mode of the following dataset: 22, 35, 22, 37, 25, 26, 31, 35, 29

The most frequently appearing values are 22 and 35, tied at two appearances apiece. Both 22 and 35 are the mode.

Range

The **range** is a measure of variability that is calculated by subtracting the smallest value from the largest value in a dataset.

Example: What is the range of the following data?

$$42, 35, 57, 28, 87, 63, 54$$

The smallest value is 28. The largest value is 87. The range is $87 - 28 = 59$.

Variance and Standard Deviation

Variance and **standard deviation** are measures of how widely or narrowly the data are spread around the mean. A low variance or standard deviation suggests that the data are clustered around a center. A high variance or standard deviation shows that the data area spread out more widely.

To calculate the variance and standard deviation:
1. Calculate the mean by adding all the data items and dividing by the number of items.
2. Find the difference between each data item and the mean.
3. Find the square of each difference.
4. Add up all the squared differences.
5. Divide the total of the squared differences by the number of data items. That is the variance.
6. Take the square root of the variance. That is the standard deviation.

Example: Calculate the variance and standard deviation for the following data set:

$$\{3, 3, 5, 7, 8, 8, 8, 10, 12, 21\}.$$

The mean is $\dfrac{3+3+5+7+8+8+8+10+12+21}{10} = \dfrac{85}{10} = 8.5$

The difference of each item x from the mean is $|8.5 - x|$. The 10 differences are:

$$5.5, 5.5, 3.5, 1.5, 0.5, 0.5, 0.5, 1.5, 3.5, 12.5$$

The squares of those ten differences are

$$30.25, 30.25, 12.25, 2.25, 0.25,$$
$$0.25, 0.25, 2.25, 12.25, 156.25$$

The sum of the ten squared differences is

$$30.25 + 30.25 + 12.25 + 2.25 + 0.25 + 0.25$$
$$+ 0.25 + 2.25 + 12.25 + 156.25 = 246.5$$

The mean of the ten squared differences is

$$\dfrac{30.25+30.25+12.25+2.25+0.25+0.25+0.25+2.25+12.25+156.25}{10} = \dfrac{246.5}{10} = 24.65$$

The variance, written as σ_2, is 24.65.

The standard deviation σ is the square root of the variance. $\sigma = \sqrt{24.65} \approx 4.96$

Displaying Statistical Data

The data obtained from sampling may be categorical (e.g., yes or no responses) or numerical. In both cases, results are displayed using a variety of graphical techniques. Geographical data is often displayed superimposed on maps.

A **histogram** is the most common form of graphical display used for numerical data obtained from random sampling is the histogram. A histogram displays data by dividing the data values into equal-sized ranges, called **bins**, and displaying bars whose height indicates the number of data items that fall into that bin. A trend line or curve can be superposed on a histogram to observe the general shape of the distribution.

If the data set is large, it may be expressed in compact form as a frequency distribution. The number of occurrences of each data point is the frequency of that value. The relative frequency is defined as the frequency divided by the total number of data points. Since the sum of the frequencies equals the number of data points, the relative frequencies add up to 1. The relative frequency of a data point, therefore, represents the probability of occurrence of that value. Thus, a distribution consisting of relative frequencies is known as a probability distribution.

The cumulative frequency of a data point is the sum of the frequencies from the beginning up to that point. A histogram is used to display a discrete frequency distribution graphically. It shows the counts of data in different ranges, the center of the data set, the spread of the data, and whether there are any outliers. It also shows whether the data has a single mode or more than one.

Example: The table below shows the summary of some test results, where people scored points ranging from 0 to 45. The total range of points has been divided into bins 0–5, 6–10, 11–15, and so on. The frequency for the first bin (labeled 5) is the number of people who scored points ranging from 0 to 5; the frequency for the second bin (labeled 10) is the number of people who scored points ranging from 6 to 10; and so on.

Points	Frequency	Cumulative Frequency	Relative Frequency
5	1	1	0.009
10	4	5	0.035
15	12	17	0.105
20	22	39	0.193
25	30	69	0.263
30	25	94	0.219
35	13	107	0.114
40	6	113	0.053
45	1	114	0.009

The histogram of the probability distribution is given below:

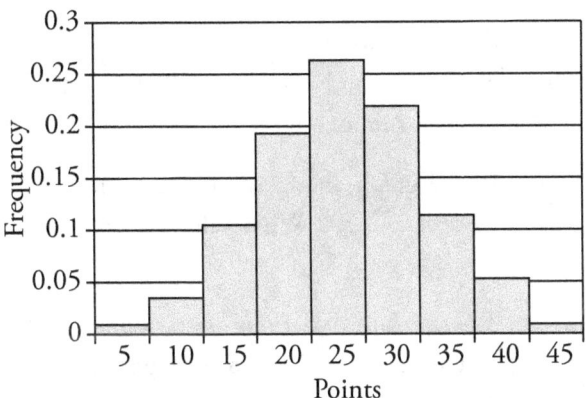

The probability distribution can be used to calculate the probability of a particular test score occurring in a certain range. For instance, the probability of a test score lying between 15 and 30 is given by the sum of the areas (assuming width of 1) of the three middle bins in the histogram above: 0.193 +1 0.263 + 0.219 = 0.675

A histogram is a **discrete frequency distribution**. It can be used to represent discrete as well as continuous data (data that can take on a continuous range of values, e.g. height) sorted in bins. A large data set of continuous data may also be represented using a **continuous frequency distribution**, which is essentially a histogram with very narrow bars. Below, a trend line has been added to the example histogram above. Notice that this approximates the most common continuous distribution, a normal or bell curve.

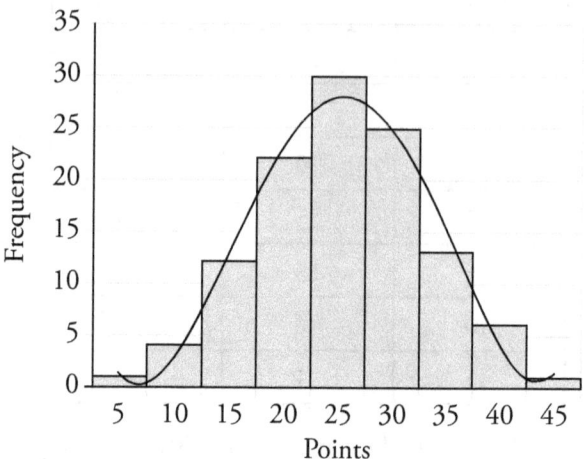

A **normal distribution** is symmetric with the mean equal to the median. The tails of the curve in both directions fall off rapidly. The spread of data is measured by the standard deviation.

Example: In the graph above, suppose the mean is 23 with a standard deviation of ±4. What values are 3 standard deviations away from the mean?

The standard deviations are measured to the right and left. So 3 standard deviations to the right is 23 + 3(4) = 35 while 3 standard deviations to the left is at 23 + 3(−4) = 11

Bar graphs are used to compare various quantities using bars of different lengths.

Example: A class had the following grades: 4 A's, 9 B's, 8 C's, 1 D, 3 F's. Graph these on a bar graph.

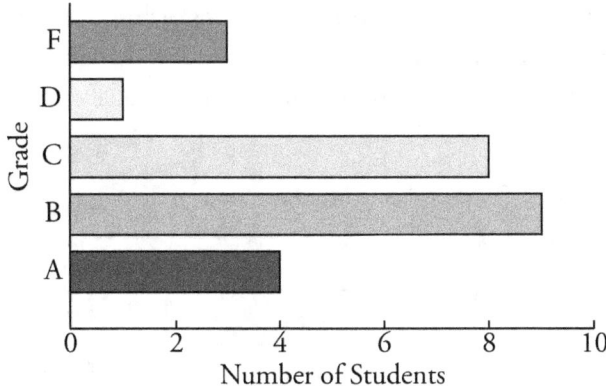

Line graphs are used to show trends, often over a period of time.

Example: Graph the following information using a line graph.

Number of National Merit Finalists/School Year						
School	90–91	91–92	92–93	93–94	94–95	95–96
Central	3	5	1	4	6	8
Wilson	4	2	3	2	3	2

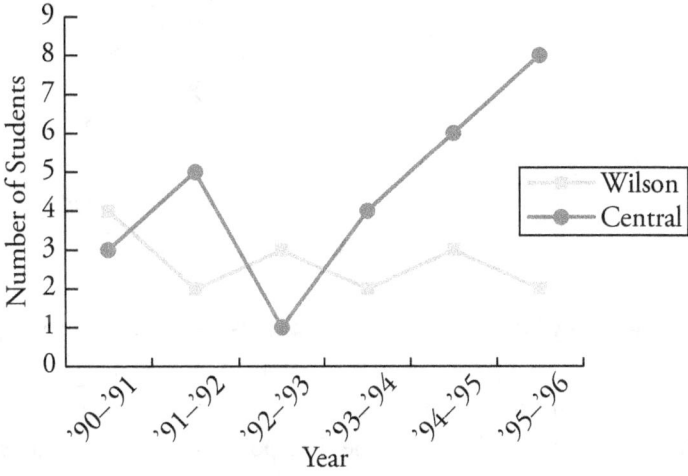

Circle graphs or **pie charts** show the relationships of various parts of a data set to each other and to the whole. Each part is shown as a percentage of the total and occupies a proportional sector of the circular area. To make a circle graph, total all the information that is to be included on the graph.

Determine the central angle to be used for each sector of the graph using the following formula: % of whole × 360°=central angle of sector Lay out the respective central angles of the various sectors, label each section and include its percentage.

Example: Graph this information on a circle graph:

Monthly Expenses	
Rent	$400
Food	$150
Utilities	$ 75
Clothes	$ 75
Church	$100
Misc.	$200

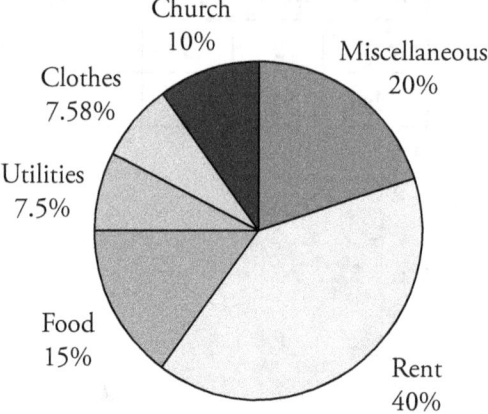

Scatter plots compare two characteristics of the same group of things or people and usually consist of a large body of data. They show how much one variable is affected by another. The relationship between the two variables is their correlation. The closer the data points come to making a straight line when plotted, the closer the correlation.

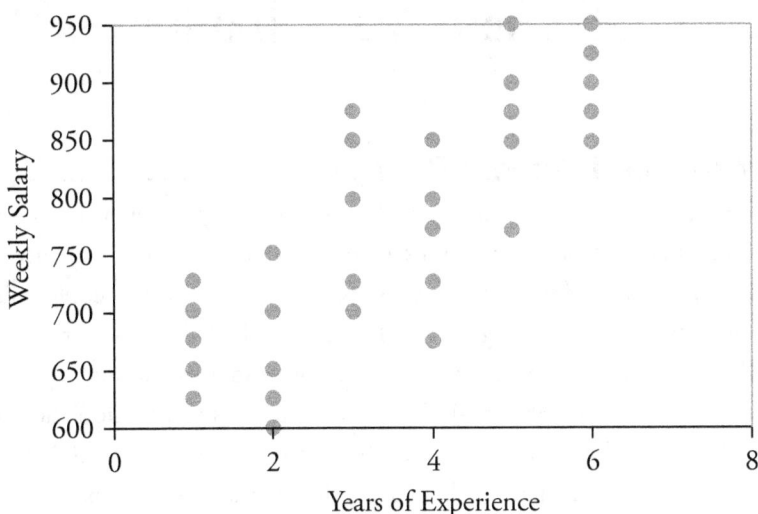

Financial Mathematics

Percents and Percent Change

The word "percent" means "per 100" and is a way to quantify sizes and make comparisons between groups of different magnitudes. For instance, if 22% of a group of 100 contestants are to be finalists, 22 people will be chosen. However, 22% of a group of 41 means only 9 will be selected.

Be aware that percent concepts can be presented in various, equivalent formats. For example, $38\% = 0.38 = \frac{38}{100}$. Cautious observation should be used with decimal places, for instance, $0.01 = 1\%$ but $0.01\% = 0.0001$.

Basic percent calculations can be performed as proportions, or linear equations.

Example: Proportion

5 is what percent of 20?

The structure of the proportion is $\frac{\text{part}}{\text{whole}} = \frac{p}{100}$

$$\frac{5}{20} = \frac{p}{100}$$
$$20p = 500$$
$$p = 25$$

The answer is 5 is 25% of 20.

Example: Linear Equation

There are 64 dogs in the kennel, 48 are collies. What percent are collies?

Restate the problem. 48 is what percent of 64?
Write an equation. $48 = n \times 64$
Solve. $\frac{48}{64} = n$

$$n = \frac{3}{4} = 0.75 = 75\%$$

75% of the dogs are collies.

Note: in the proportion method, the percent value is used directly, while in the linear equation method, the percent value must be converted to its decimal equivalent.

Example: The auditorium was filled to 90% capacity. There were 558 seats occupied. What is the capacity of the auditorium?

Restate the problem. 90% of what number is 558?
Write an equation. $0.9n = 558$
Solve. $n = \dfrac{558}{.9}$
$n = 620$

The capacity of the auditorium is 620 people.

Example: A pair of shoes costs $42.00. Sales tax is 6%. What is the total cost of the shoes?

Restate the problem. What is 6% of 42?
Write an equation. $n = 0.06 \times 42$
Solve. $n = 2.52$

Add the sales tax to the cost. $42.00 + $2.52 = $44.52

The total cost of the shoes, including sales tax, is $44.52.

Example: A share of stock in Wonder Widgets cost $52 at the start of the day and had risen to $83 at the close of trading. What was the percent change in the share price?

First find the actual change in the share price: 83 − 52 = $31 change in share price.

When considering **percent change**, the change should be compared to the original. In this case, consider "31 is what percent of 52?" as 52 represents the original share price.

Solving by linear equation:

$$31 = r \times 52$$
$$r = \dfrac{31}{52} \approx .60 = 60\%$$

Financial Mathematics   **81**

Example: A sweater priced at $47 is marked down 15%. What is the new price, to the nearest dollar, of the sweater?

Find: 15% of 47 = d, where d represents by how many dollars the price is lowered.

$$0.15\,(47) = 7.05 \text{ dollars decrease}$$

Thus the new price of the sweater will be $47 - 7.05 \approx \$40$.

APR and Interest Rate Calculations

The financial world relies heavily on the percent calculation, and numerous instances involve calculations that are more complicated, and often more repetitive than the basic examples shown above.

Often, for instance, a bank will charge an **annual percentage rate**, or **APR**, of $p\%$ on a loan. In the most basic sense, if a consumer borrowed $100 for a year with an APR of 9%, he or she would owe the bank $109 upon repayment. But this assumes the interest is charged annually, or only one time during the year.

In reality, lending institutions calculate interest at intervals throughout the year. Interest can be charged, for example, quarterly (4 times a year), or daily (365 times a year), or even continuously. This periodic rate affects the actual interest paid on the loan (or, conversely, earned on an investment).

A valuable calculation used to understand the reality of interest rate charges is the **effective interest rate** formula: $E = \left(1 + \frac{r}{n}\right)^n - 1$ where r represents the APR, and n is the number of times the interest is calculated in a year.

Example: A credit card offers an APR of 18%. Charges are calculated monthly. Find the effective interest rate on the card.

$$E = \left(1 + \frac{r}{n}\right)^n - 1 = \left(1 + \frac{.18}{12}\right)^{12} - 1 \approx .196 = 19.6\%$$

This calculation reminds the consumer that interest charges are, in fact, higher than the straight 18% offered with the card.

The effective interest rate formula can be expanded, so to speak, to become the **compound interest formula**: $A = P\left(1 + \frac{r}{n}\right)^{nt}$. This calculation is most commonly used for investments lasting a year or more when interest is awarded at multiple intervals throughout the year. P stands for the principal, or the initial investment amount. The values for r and n are the same as in the effective interest formula, and t represents the number of years that the money is invested. The calculation yields the final balance, A.

Example: Suppose a person invests $2,000 in an account earning an annual interest rate of 3.3% compounded quarterly. How much money will be in the account at the end of 5 years?

$$A = 2,000\left(1 + \frac{.033}{4}\right)^{4 \cdot 5} = 2,000(1.00825)^{20} = \$2,357.19$$

As mentioned earlier, a periodic interest rate can be measured in various intervals from months, to days, to seconds, and even to infinitely small intervals, which results in continuous compounding. To calculate interest **compounded continuously**, a new formula must be used: $A = Pe^{rt}$ where e is the constant base of the natural log function. (e is an irrational number and has an approximate value of 2.718)

Example: Find the balance on an account starting with an initial deposit of $750 after 10 years of interest compounded continuously at an annual rate of 4.9%.

Using the continuously compounded interest formula:

$$A = Pe^{rt} = 750e^{.049(10)} = \$1,224.24$$

Present and Future Value

The concept of **present and future value** is used to present the idea that money may be worth more (or less) later, when considering its investment potential, or that expenses may be greater as time goes by due to inflation. For instance, in the previous example, the approximate future value of the $750 (present value) under the given investment conditions is $1,224.

Example: Find the present value needed to have, in 18 years, a future value of $25,000 with an investment rate of 5% compounded monthly.

To determine this amount, the compound interest formula can be used with the future value representing A:

$$25{,}000 = P\left(1 + \frac{.05}{12}\right)^{12(18)}$$
$$25{,}000 = P(2.455)$$
$$P \approx \$10{,}183$$

In other words, a present value investment of a little over $10,000 can more than double to $25,000 in 18 years, given these investment conditions.

Example: If the rate of inflation averages 2.4%, what will be the future cost, in 5 years, of a gallon of milk with a present value of $3.50?

Inflation calculations are performed with the continuously compounding formula.

$$A = Pe^{rt} = 3.5e^{(.024)(5)} = \$3.95$$

Geometry

Polygons: Triangles, Quadrilaterals, and More

Polygons, simple closed **two-dimensional figures** composed of line segments, are named according to the number of sides they have.

A **triangle** *is a polygon with three sides.*

The sum of the measures of the angles of a triangle is 180o.

Triangles can be classified by the types of angles or the lengths of their sides.

An **acute** triangle has exactly three *acute* angles.

A **right** triangle has one *right* angle.

An **obtuse** triangle has one *obtuse* angle.

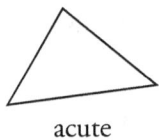
acute

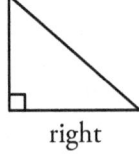
right

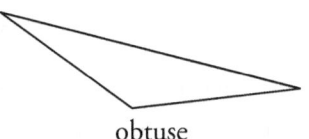
obtuse

All *three* sides of an **equilateral** triangle are the same length.

Two sides of an **isosceles** triangle are the same length.

None of the sides of a **scalene** triangle are the same length.

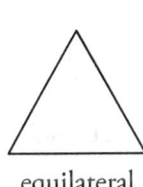
equilateral

isosceles

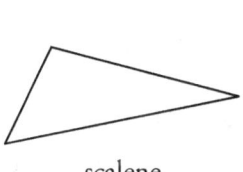
scalene

Additionally, an equilateral triangle has 3 congruent angles and an isosceles triangle has two congruent base angles. (The congruent angles sit at each end of the non-congruent side of the triangle)

Example: Can a triangle have two right angles? No. A right angle measures 90°, therefore the sum of two right angles would be 180° and there could not be third angle.

Example: Can a triangle have two obtuse angles? No. Since an obtuse angle measures more than 90° the sum of two obtuse angles would be greater than 180°.

Example: Given equilateral triangle TRY, find m∠R and length of side RY.

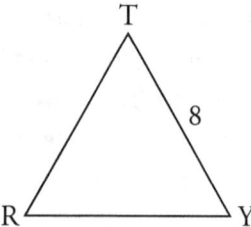

Since the triangle is equilateral, every side is the same length. So RY must also be 8 units long. Also, an equilateral triangle has 3 congruent angles. If the sum of the angles of a triangle is 180°, then each angle measures 180 ÷ 3 = 60. Therefore m∠R = 60°.

A **quadrilateral** *is a polygon with four sides.*
The sum of the measures of the angles of a quadrilateral is 360°.

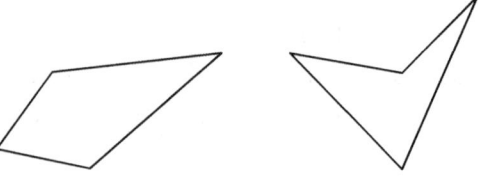

Certain quadrilaterals are specifically named according to their unique properties.

A **trapezoid** is a quadrilateral with exactly *one* pair of parallel sides.

The two parallel sides of a trapezoid are called the bases, and the two non-parallel sides are called the legs. If the two legs are the same length, then the trapezoid is called isosceles.

The segment connecting the two midpoints of the legs is called the median. The median has the following two properties:
1. The median is parallel to the two bases.
2. The length of the median is equal to one-half the sum of the length of the two bases.

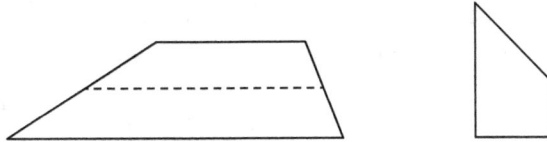

In an **isosceles trapezoid**, the non-parallel sides are congruent.

An isosceles trapezoid has the following properties:
1. The diagonals of an isosceles trapezoid are congruent.
2. The base angles of an isosceles trapezoid are congruent.

Example: An isosceles trapezoid has a diagonal of 10 and a base angle measure of 30°. Find the measure of the other 3 angles.

Based on the properties of trapezoids, the measure of the other base angle is 30° and the measure of the other diagonal is 10. The other two angles have a measure of

$$360 = 30(2) + 2x$$
$$x = 150°$$

The other two angles measure 150° each.

A **parallelogram** is a quadrilateral with *two* pairs of parallel sides and has the following properties:

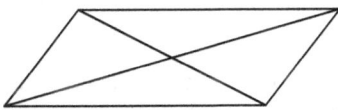

1. The diagonals bisect each other.
2. Each diagonal divides the parallelogram into two congruent triangles.
3. Both pairs of opposite sides are congruent.
4. Both pairs of opposite angles are congruent.
5. Two adjacent angles are supplementary.

Example: Find the measures of the other three angles of a parallelogram if one angle measures 38°.

Since opposite angles are equal, there are two angles measuring 38°. Since adjacent angles are supplementary, 180 − 38 = 142. Hence the other two angles measure 142° each.

Example: The measures of two adjacent angles of a parallelogram are $3x + 40$ and $x + 70$. Find the measures of each angle.

$$2(3x+40)+2(x+70)=360$$
$$6x+80+2x+140=360$$
$$8x+220=360$$
$$8x=140$$
$$x=17.5$$
$$3x+40=92.5$$
$$x+70=87.5$$

Thus the angles measure 92.5°, 92.5°, 87.5°, and 87.5°.

A **rectangle** is a parallelogram with a right angle. Since a rectangle is a special type of parallelogram, it exhibits all the properties of a parallelogram. All the angles of a rectangle are right angles because of congruent opposite angles. Additionally, the diagonals of a rectangle are congruent.

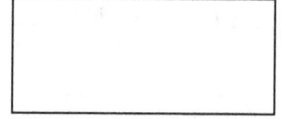

A **rhombus** is a parallelogram with all sides equal in length. A rhombus also has all the properties of a parallelogram. Additionally, its diagonals are perpendicular to each other and they bisect its angles.

Example: In rhombus $ABCD$ side $AB = 3x - 7$ and side $CD = x + 15$. Find the length of each side.

Since all the sides are the same length, $3x - 7 = x + 15$
$$2x = 22$$
$$x = 11$$

Since $3(11) - 7 = 25$ and $11 + 15 = 25$, each side measures 25 units.

A **square** is a rectangle with all sides equal in length. A **square** has all the properties of a rectangle *and* a rhombus.

Example: True or false?

All squares are rhombuses. **True**

All parallelograms are rectangles. **False — *some* parallelograms are rectangles**

All rectangles are parallelograms. **True**

Some rhombuses are squares. **True**

Some rectangles are trapezoids. **False — only *one* pair of parallel sides**

All quadrilaterals are parallelograms. **False — some quadrilaterals are parallelograms**

Some squares are rectangles. **False — all squares are rectangles**

Some parallelograms are rhombuses. **True**

Polygons can have an infinite number of sides, but some of the most commonly referenced, beyond triangle and quadrilateral are as follows:

5 sides	pentagon
6 sides	hexagon
8 sides	octagon
10 sides	decagon

Applications of Polygons: Perimeter, Area, Congruence, Similarity, and Pythagorean Theorem —

The **perimeter** of any polygon is the sum of the lengths of the sides.
The **area** of a polygon is the number of square units covered by the figure.

Figure	Area Formula	Perimeter Formula
Rectangle	LW	$2(L + W)$
Triangle	$\frac{1}{2}bh$	$a + b + c$
Parallelogram	bh	sum of lengths of sides
Trapezoid	$\frac{1}{2}h(a+b)$	sum of lengths of sides

Example: A farmer has a piece of land shaped as shown below. He wishes to fence this land at an estimated cost of $25 per linear foot. What is the total cost of fencing this property to the nearest foot.

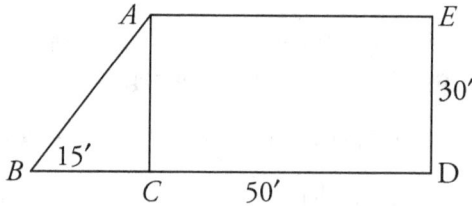

From the right triangle ABC, AC = 30 and BC = 15. (see page 100 for information about the Pythagorean Theorem)

Since $(AB) = (AC)^2 + (BC)^2$
$(AB) = (30)^2 + (15)^2$

So $\sqrt{(AB)^2} = AB = \sqrt{1125} = 33.5410$ feet

To the nearest foot AB = 34 feet.

Perimeter of the piece of land is $AB + BC + CD + DE + EA$ = 34 + 15 + 50 + 30 + 50 = 179 feet

Cost of fencing = \$25 × 179 = \$4,475.00

Example: Find the area of a parallelogram whose base is 6.5 cm and the height of the altitude to that base is 3.7 cm.

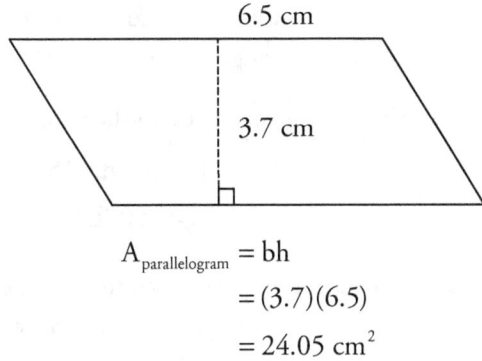

$$A_{parallelogram} = bh$$
$$= (3.7)(6.5)$$
$$= 24.05 \text{ cm}^2$$

Example: Find the area of this triangle.

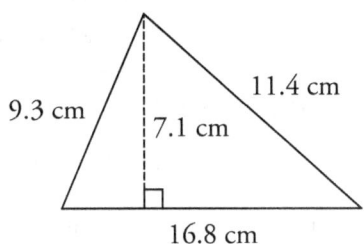

$$A_{triangle} = \frac{1}{2}bh$$
$$= 0.5(16.8)(7.1)$$
$$= 59.64 \text{ cm}^2$$

Example: What will be the cost of carpeting a rectangular office that measures 12 feet by 15 feet if the carpet costs $12.50 per square yard?

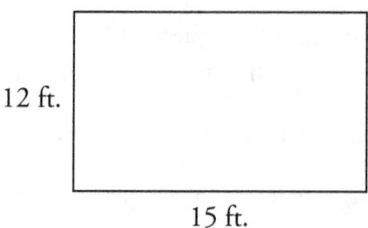

12 ft.

15 ft.

The problem is asking you to determine the area of the office. The area of a rectangle is *length* × *width* = *A*

Substitute the given values in the equation $A = lw$

$$A = (12 \text{ ft.})(15 \text{ ft.})$$
$$A = 180 \text{ ft.}^2$$

The problem asked you to determine the cost of carpet at $12.50 per square yard. First, you need to convert 180 ft.² into yards².

$$1 \text{ yd.} = 3 \text{ ft.}$$
$$(1 \text{ yard})(1 \text{ yard}) = (3 \text{ feet})(3 \text{ feet})$$
$$1 \text{ yd}^2 = 9 \text{ ft}^2$$

Hence, 180 ft² 1 yd² 20

$$1 = 9 \text{ ft}^2 = 1 = 20 \text{ yd}^2$$

The carpet cost $12.50 per square yard; thus the cost of carpeting the office described is $12.50 × 20 = $250.00.

Congruence

Congruent figures have the same size and shape. If one is placed above the other, it will fit exactly. Congruent lines have the same length. Congruent angles have equal measures. The symbol for congruent is ≅. Polygons (pentagons) *ABCDE* and *VWXYZ* are congruent. They are exactly the same size and shape.

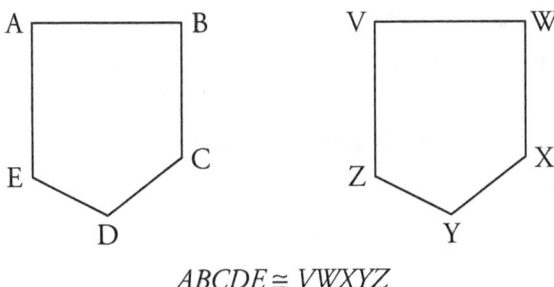

$ABCDE \cong VWXYZ$

Corresponding parts are those congruent angles and congruent sides, that is:

Corresponding Angles	Corresponding Sides
$\angle A \leftrightarrow \angle V$	$AB \leftrightarrow VW$
$\angle B \leftrightarrow \angle W$	$BC \leftrightarrow WX$
$\angle C \leftrightarrow \angle X$	$CD \leftrightarrow XY$
$\angle D \leftrightarrow \angle Y$	$DE \leftrightarrow YZ$
$\angle E \leftrightarrow \angle Z$	$AE \leftrightarrow VZ$

To prove triangles congruent, it is not always necessary to demonstrate that all corresponding sides and angles are congruent. There are several "shortcut" methods described below.

- The **SAS Postulate** (side-angle-side) states that if two sides and the included angle of one triangle are congruent to two sides and the included angle of another triangle, then the two triangles are congruent.

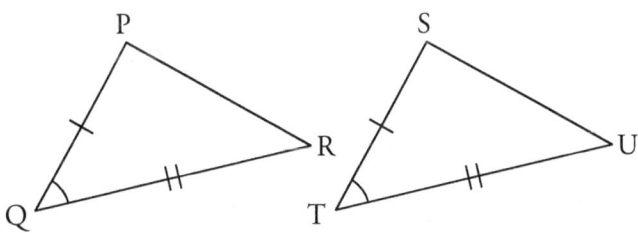

To see why this is true, imagine moving the triangle PQR (shown above) in such a way that the point P coincides with the point S, and line segment PQ coincides with line segment ST. Point Q will then coincide with T since

PQ ≅ ST. Also, segment QR will coincide with TU, because ∠Q ≅ ∠T. Point R will coincide with U, because QR ≅ TU. Since P and S coincide and R and U coincide, line PR will coincide with SU because two lines cannot enclose a space. Thus the two triangles match perfectly point for point and are congruent.

Example: Are the following triangles congruent?

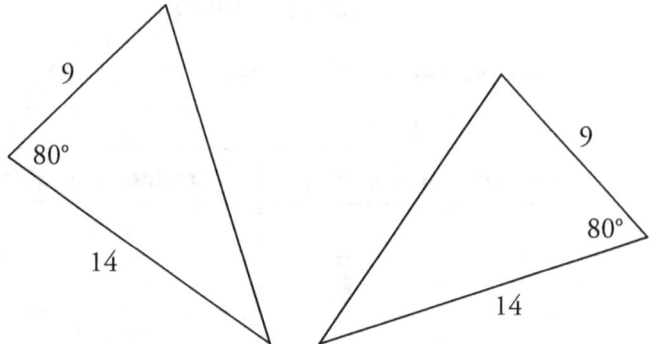

Each of the two triangles has a side that is 14 units and another that is 9 units. The angle included in the sides is 80° in both triangles. Therefore, the triangles are congruent by SAS.

- The **SSS Postulate** (side-side-side) states that if three sides of one triangle are congruent to three sides of another triangle, then the two triangles are congruent.

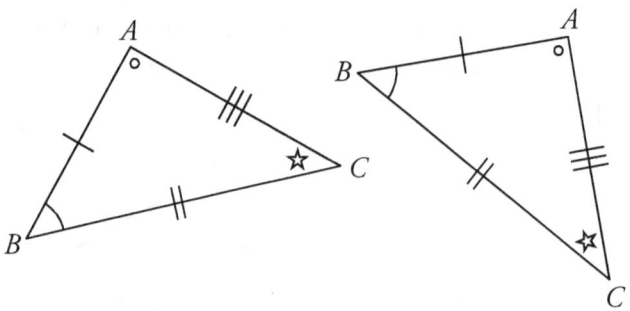

Since $AB \cong XY$, $BC \cong YZ$ and $AC \cong XZ$, then $\triangle ABC \cong \triangle XYZ$.

Example: Given isosceles triangle *ABC* with *D* being the midpoint of base *AC*, prove that the two triangles *ABD* and *ADC* are congruent.

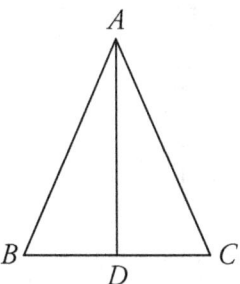

Proof:

1. Isosceles triangle *ABC*, *D* midpoint of base *AC* — Given
2. $AB \cong AC$ — An isosceles triangle has two congruent sides
3. $BD \cong DC$ — Midpoint divides a line into two equal parts
4. $AD \cong AD$ — Reflexive property
5. $\triangle ABD \cong \triangle BCD$ — SSS

- The **ASA Postulate** (angle-side-angle) states that if two angles and the included side of one triangle are congruent to two angles and the included side of another triangle, the triangles are congruent.

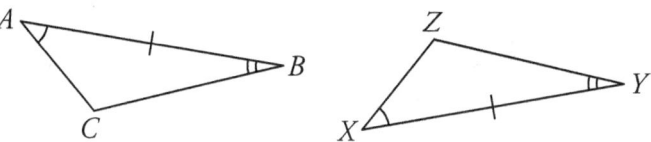

$\angle A \cong \angle X$, $\angle B \cong \angle Y$, $AB \cong XY$ then $\triangle ABC \cong \triangle XYZ$ by ASA

Example: Given two right triangles with right angles at B and L and with one leg (*AB* and *KL*) of each measuring 6 cm and the adjacent angle 37°, prove the triangles are congruent.

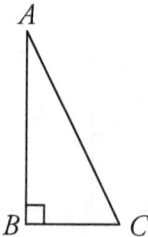 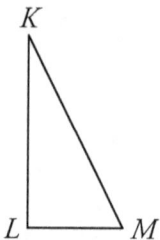

Proof:

1. Right △*ABC* and △*KLM*   Given
   AB = *KL* = 6 cm
   ∠*A* = ∠*K* = 37°
2. *AB* ≅ *KL*   Figures with the same
   ∠*A* ≅ ∠*K*   measure are congruent
3. ∠*B* ≅ ∠*L*   All right angles are congruent.
4. △*ABC* ≅ △*KLM*   ASA

Example: What method could be used to prove that triangles *ABC* and *ADE* are congruent?

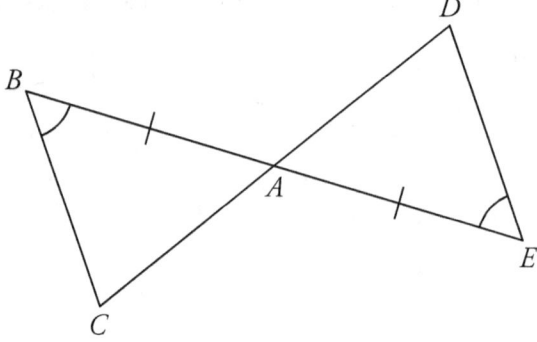

The sides *AB* and *AE* are given as congruent, as are ∠*BAC* and ∠*DAE*. ∠*BAC* and ∠*DAE* are vertical angles and are therefore congruent. Thus triangles △*ABC* and △*ADE* are congruent by the ASA postulate.

Similarity

Two figures that have the same shape are **similar**. Polygons are similar if and only if corresponding angles are congruent and corresponding sides are in proportion. Corresponding parts of similar polygons are proportional.

Example: Given two similar quadrilaterals. Find the lengths of sides x, y, and z.

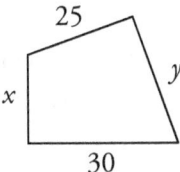 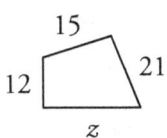

Since corresponding sides are proportional:

= so the scale is

$$\frac{12}{x} = \frac{3}{5} \qquad \frac{21}{y} = \frac{3}{5} \qquad \frac{z}{30} = \frac{3}{5}$$
$$3x = 60 \qquad 3y = 105 \qquad 5z = 90$$
$$x = 20 \qquad y = 35 \qquad z = 18$$

Example: Given the rectangles below, compare the area and perimeter.

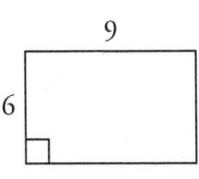 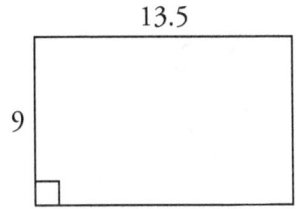

$A = LW$	$A = LW$	1. write formula
$A = (6)(9)$	$A = (9)(13.5)$	2. substitute known values
$A = 54$ sq. units	$A = 121.5$ sq. units	3. compute
$P = 2(L + W)$	$P = 2(L + W)$	1. write formula
$P = 2(6 + 9)$	$P = 2(9 + 13.5)$	2. substitute known values
$P = 30$ units	$P = 45$ units	3. compute

Geometry

Notice that the areas relate to each other in the following manner:

Ratio of sides 9/13.5 = 2/3

Multiply the first area by the square of the reciprocal $(3/2)^2$ to get the second area. $54 \times (3/2)^2 = 121.5$

The perimeters relate to each other in the following manner: Ratio of sides: 9/13.5 = 2/3

Multiply the perimeter of the first by the reciprocal of the ratio to get the perimeter of the second $30 \times 3/2 = 45$

Just as for congruence, there are shortcut methods that can be used to prove triangle similarity.

- According to the **AA Similarity Postulate**, if two angles of one triangle are congruent to two angles of another triangle, then the triangles are similar. It is obvious that if two of the corresponding angles are congruent, the third set of corresponding angles must be congruent as well. Hence, showing AA is sufficient to prove that two triangles are similar.
- The **SAS Similarity Theorem** states that, if an angle of one triangle is congruent to an angle of another triangle and the sides adjacent to those angles are in proportion, then the triangles are similar.

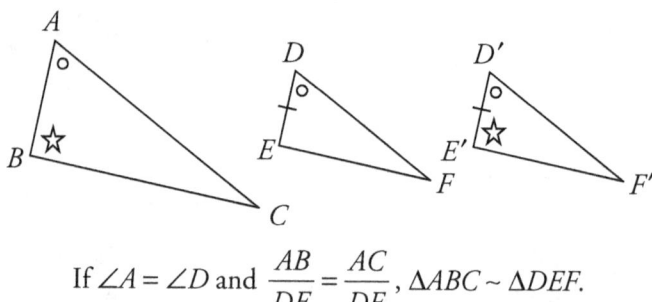

If $\angle A = \angle D$ and $\dfrac{AB}{DE} = \dfrac{AC}{DF}$, $\triangle ABC \sim \triangle DEF$.

Example: A graphic artist is designing a logo containing two triangles. The artist wants the triangles to be similar. Determine whether the artist has created similar triangles.

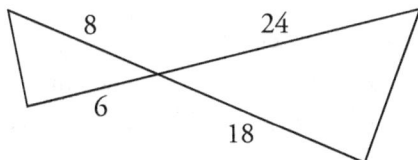

The sides are proportional (8/24 = 6/18 = 1/3) and vertical angles are congruent. The two triangles are therefore similar by the SAS similarity theorem.

- According to the **SSS Similarity Theorem**, if the sides of two triangles are in proportion, then the triangles are similar.

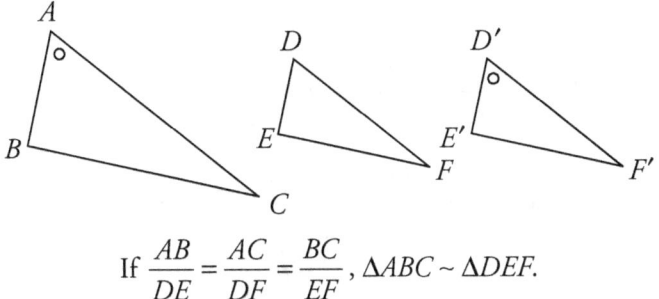

If $\dfrac{AB}{DE} = \dfrac{AC}{DF} = \dfrac{BC}{EF}$, $\triangle ABC \sim \triangle DEF$.

Example: Tommy draws and cuts out 2 triangles for a school project. One of them has sides of 3, 6, and 9 inches. The other triangle has sides of 2, 4, and 6. Is there a relationship between the two triangles?

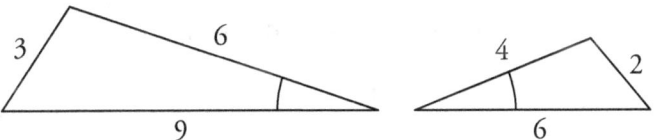

Determine the proportions of the corresponding sides.

$$\dfrac{2}{3} \qquad \dfrac{4}{6} = \dfrac{2}{3} \qquad \dfrac{6}{9} = \dfrac{2}{3}$$

The smaller triangle is 2/3 the size of the large triangle, therefore they are similar triangles by the SSS similarity theorem.

Some of the problems in the above skill section on formulas for geometric figures treat cases where dimensions of figures or solids are changed, resulting in changes to certain other parameters.

The Pythagorean Theorem

Given any right-angles triangle, $\triangle ABC$, the square of the hypotenuse is equal to the sum of the squares of the other two sides.

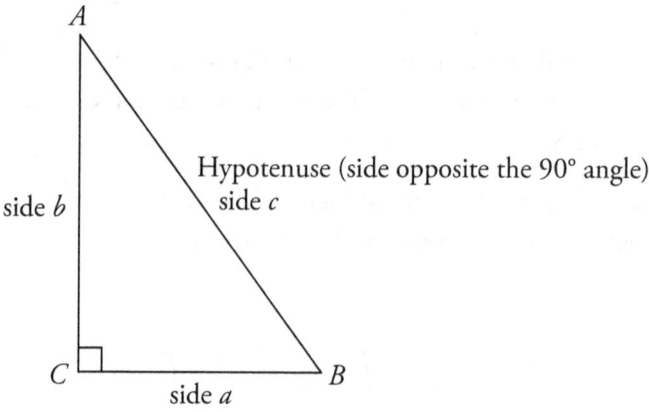

This theorem says that $(AB)^2 = (BC)^2 + (AC)^2$
Or
$c^2 = a^2 + b^2$

Example: Find the area and perimeter of a rectangle if its length is 12 inches and its diagonal is 15 inches.

1. Draw and label sketch.
2. Since the height is still needed use Pythagorean formula to find missing leg of the triangle.

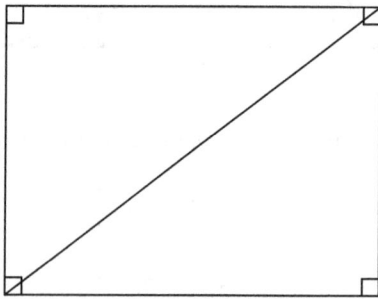

$$A^2 + B^2 = C^2$$
$$A^2 + 12^2 = 15^2$$
$$A^2 = 15^2 - 12^2$$
$$A^2 = 81$$
$$A = 9$$

Now use this information to find the area and perimeter.

$A = LW$	$P = 2(L + W)$	1.	write formula
$A = (12)(9)$	$P = 2(12 + 9)$	2.	substitute
$A = 108$ in²	$P = 42$ inches	3.	solve

Parallel and Perpendicular Lines: Properties and Applications

Parallel lines in two dimensions can be sufficiently defined as lines that do not intersect. In three dimensions, however, this definition is insufficient. **Parallel lines** in three dimensions are defined as lines for which every pair of nearest points on the lines has a fixed distance.

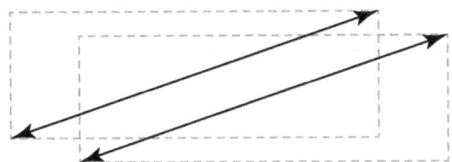

Lines in three dimensions that do not intersect and are not parallel are called **skew lines**. Parallel lines are coplanar, skew lines are not.

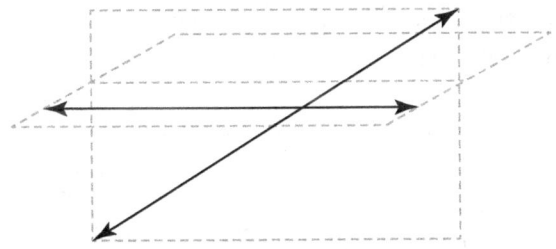

Two planes intersect on a single line. If two planes do not intersect, then they are parallel. Parallel and non-parallel planes are shown in the diagram below.

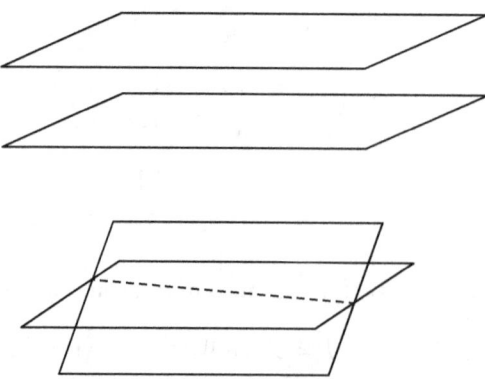

Parallelism between two planes may also be defined in the same way as parallel lines: the distance between any pair of nearest points (one point on each plane) is constant.

Perpendicularity of lines and planes in three dimensions is largely similar to that of two dimensions. Two lines are **perpendicular**, in two or three dimensions, if they intersect at a point and form 90° angles between them. Consequently, perpendicular lines are always coplanar.

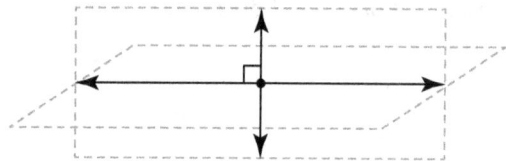

Notice that, for any line and coincident point on that line, there are an infinite number of perpendicular lines to the line through that point. In two dimensions, there is only one.

Two planes are perpendicular if they intersect and the angles formed between them are 90°. For any given plane and line on that plane, there is only one perpendicular plane.

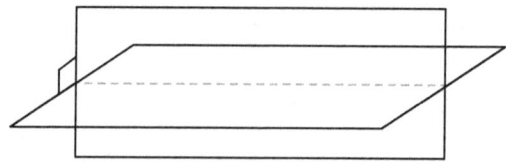

The **Parallel Postulate** in Euclidean planar geometry states that if a line l is crossed by two other lines m and n (where the crossings are not at the same point on l), then m and n intersect on the side of l where the sum of the interior angles α and β is less than 180°. This scenario is illustrated below.

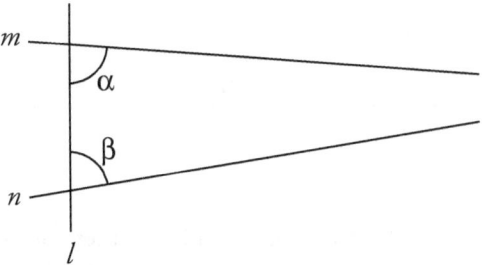

Based on this definition, a number of implications and equivalent formulations can be derived. First, note that the lines m and n intersect on the right-hand side of l above only if α + β < 180°. This implies that if α and β are both 90° and, therefore, α + β = 180°, then the lines do not intersect on either side. This is illustrated below.

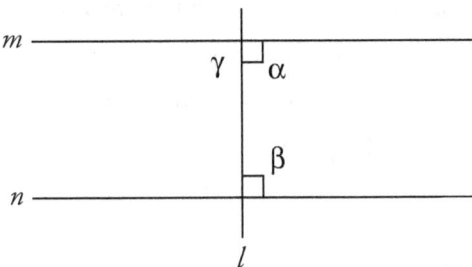

The supplementary angles formed by the intersection of l and m (and the intersection of l and n) must sum to 180°:

$$\alpha + \gamma = 180° \qquad \beta + \delta = 180°$$

Since these sums are both equal to 180°, the lines m and n do not intersect on either side of l. That is to say, these lines are **parallel**.

Let the non-intersecting lines m and n used in the above discussion remain parallel, but adjust l such that the interior angles are no longer right angles.

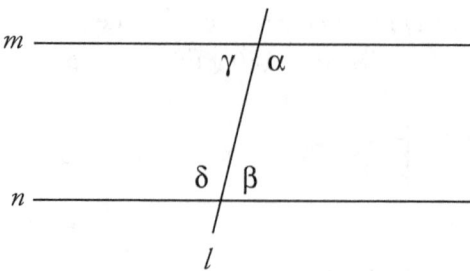

The Parallel Postulate still applies, and it is therefore still the case that $\alpha + \beta = 180°$ and $\gamma + \delta = 180°$. Combined with the fact that $\alpha + \gamma = 180°$ and $\beta + \delta = 180°$, the **Alternate Interior Angle Theorem** can be justified. This theorem states that if two parallel lines are cut by a transversal, the alternate interior angles are congruent.

By manipulating the four relations based on the above diagram, the relationships between alternate interior angles (γ and β form one set of alternate interior angles, and α and δ form the other) can be established.

$$\alpha = 180° - \beta$$
$$\alpha + \gamma = 180° = 180° - \beta + \gamma$$
$$-\beta + \gamma = 0$$
$$\gamma = \beta$$

By the same reasoning,

$$\gamma = 180° - \delta$$
$$\beta + \delta = 180° = \beta + 180° - \delta$$
$$\beta = \delta$$

One of the consequences of the Parallel Postulate, in addition the Alternate Interior Angle Theorem, is that **corresponding angles** are equal. If two parallel lines are cut by a transversal line, then the corresponding angles are equal. The diagram below illustrates one set of corresponding angles (α and β) for the parallel lines m and n cut by l.

That α and β are equal can be proven as follows.

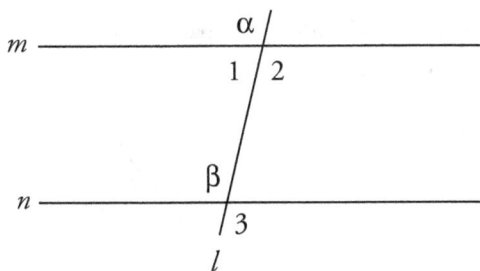

∠β = ∠2   Alternate Interior Angle Theorem
∠1 + ∠2 = 180°   Supplementary angles
∠2 = 180° − ∠1
∠1 + ∠α = 180°   Supplementary angles
∠α = 180° − ∠1
∠2 = 180° − ∠1 = ∠α
∠2 = ∠α
∠β = ∠2 = ∠α
∠β = ∠α

Thus, it has been proven that corresponding angles are equal.

Note, also, that the above proof also demonstrates that vertical angles are equal (∠2 = ∠α). Thus, opposite angles formed by the intersection of two lines (called **vertical angles**) are equal. Furthermore, **alternate exterior angles** (angles α and 1 in the diagram above) are also equal.

∠β = ∠3   Vertical angles
∠α = ∠2   Vertical angles
∠β = ∠2   Alternate Interior Angle Theorem
∠α = ∠2 = ∠β = ∠3
∠α = ∠3

Example: In the diagram below, ℓ_1 is parallel to ℓ_2 and $m\angle a = 55°$. Find $m\angle b$ and $m\angle c$.

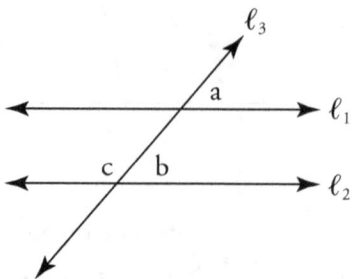

With ℓ_3 as the transversal through 2 parallel lines, $\angle a$ and $\angle b$ are congruent corresponding angles. Therefore $m\angle b = 55°$. Since $\angle c$ and $\angle b$ form a linear pair, they are supplementary: their measures add to 180°. $180 - 55 = 125 = m\angle c$.

Circles: Properties, Arcs, Angles, and Applications

A circle is defined as the set of all points in a plane that are equidistant from a given center. The distance from the center to the circle is called the **radius**. The length of two radii, or the distance across the entire circle, measured through the center, is the **diameter**.

The distance around a circle is the **circumference**. The ratio of the circumference to the diameter is represented by the Greek letter pi.

$$\pi \sim 3.14 \sim \frac{22}{7}$$

The circumference of a circle is found by the formula $C = 2\pi r$ or $C = \pi d$ where r is the radius of the circle and d is the diameter.

The **area** of a circle is found by the formula $A = \pi r^2$.

Example: Find the circumference and area of a circle whose radius is 7 meters.

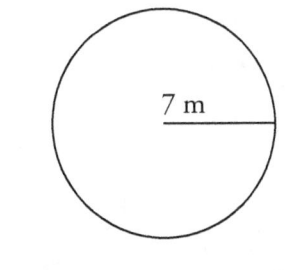

$C = 2\pi r$ $A = \pi r^2$
$= 2(3.14)(7)$ $= 3.14(7)(7)$
$= 43.96$ m $= 153.86$ m^2

If you draw two radii in a circle, the angle they form with the center as the vertex is a **central angle**. The piece of the circle "inside" the angle is an arc. Just like a central angle, an arc can have any degree measure from 0 to 360. The measure of an arc is equal to the measure of the central angle that forms the arc. Since a diameter forms a semicircle and the measure of a straight angle like a diameter is 180°, the measure of a semicircle is also 180°.

Given two points on a circle, the two points form two different arcs. Except in the case of semicircles, one of the two arcs will always be greater than 180° and the other will be less than 180°. The arc less than 180° is a **minor arc** and the arc greater than 180° is a **major arc**.

Example: If $m\angle BAD = 45°$, what is the measure of the major arc BD?

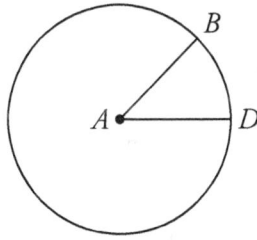

The minor arc BD is the same as $m\angle BAD = 45°$. Since the sum of the minor and major arcs formed by two points on a circle always add to 360°, the major arc BD must be 360° − 45° = 315°.

Example: If $\overline{AC}$ is a diameter of the circle below, what is the measure of ∡BDC?

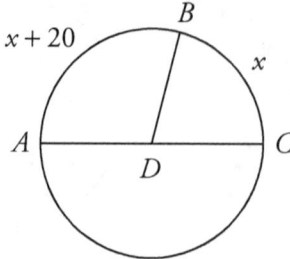

Since the diameter forms a semicircle ABC with a measure of 180°, the following expression applies.

$$(x + 20) + x = 180°$$

Solving for x yields:

$$2x + 20 = 180°$$
$$2x = 160°$$
$$x = 80°$$

Finally, since the measure of an arc is the same as the measure of the central angle, ∡BDC = x = 80°

Although an arc has a measure associated with the degree measure of the corresponding central angle, it also has an **arc length** that is a fraction of the circumference of the circle. For each central angle and its associated arc, there is a **sector** of the circle that resembles a pie piece. The area of such a sector is a fraction of the area of the circle. The fractions used for the area of a sector and length of its associated arc are both equal to the ratio of the central angle to 360°.

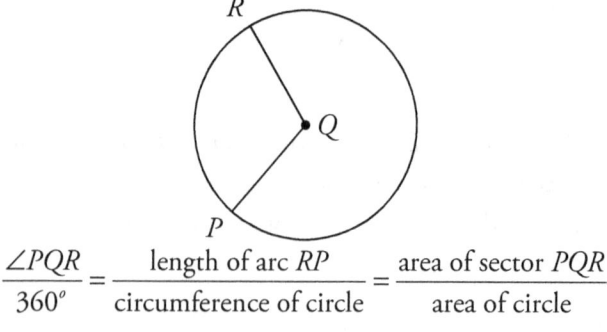

$$\frac{\angle PQR}{360°} = \frac{\text{length of arc } RP}{\text{circumference of circle}} = \frac{\text{area of sector } PQR}{\text{area of circle}}$$

108   CLEP College Level Math

Example: Circle A as a radius of 4 cm. What is the length of arc *ED*?

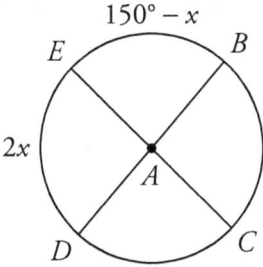

Because arcs *BE* and *ED* form a semicircle, arc *BED* is 180°. Use this to write an equation and solve for *x*.

$$(150° - x) + 2x = 180°$$
$$x = 30°$$

The angle corresponding to arc *ED* is thus $2x = 60°$. The ratio of this arc to that of the entire circle (360°) must be the same as the ratio of the arc length (labeled *L*) to the circumference of the circle, as shown below.

$$\frac{60°}{360°} = \frac{L}{2\pi(4\text{cm})}$$

$$L = \frac{1}{6} 2\pi(4\text{cm}) = \frac{4}{3}\pi \text{ cm} \approx 4.19\text{cm}$$

Example: The radius of circle M is 3 cm. The length of arc *PF* is 2π cm. What is the area of sector *MPF*?

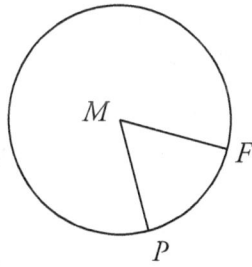

The circumference of the circle is $2\pi(3 \text{ cm})$, making the circumference equal to 6π cm. The total area of the circle is $\pi(3 \text{ cm})^2$, or 9π cm². The ratio of the arc length *PF* to the circumference of the circle must be the same as the ratio of the area of sector *MPF* (labeled *A*) to the total area of the circle. Thus,

$$\frac{A}{9\pi \text{ cm}^2} = \frac{2\pi}{6\pi}$$

$$A = \frac{1}{3} 9\pi \text{ cm}^2 = 3\pi \text{ cm}^2 \approx 9.42 \text{ cm}^2$$

An **inscribed angle** is an angle whose vertex is on the circle's edge. Such an angle could be formed by two chords, two diameters, two secants, or a secant and a tangent. An inscribed angle has one arc of the circle in its interior. The measure of the inscribed angle is one-half the measure of its intercepted arc. If two inscribed angles intercept the same arc, the two angles are congruent (i.e., their measures are equal). If an inscribed angle intercepts an entire semicircle, the angle is a right angle.

Example: Given circle *B*, and that the measure of minor arc *XY* is 100°, find m∠B and m∠P.

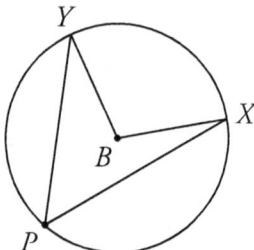

Since ∠B is a central angle, its measure is equivalent to the intercepted arc: 100°.

However, ∠P, with vertex on the edge of the circle, is an inscribed angle so its measure is half of the arc: 50°.

Logic and Sets

Logical Connectives and Quantifiers

A simple statement represents a simple idea, that can be described as either "true" or "false", but not both. A simple statement is represented by a small letter of the alphabet.

Example: "Today is Monday." This is a simple statement since it can be determine that this statement is either true or false. We can write p = "Today is Monday".

Example: "John, please be quiet". This is not considered a simple statement in our study of logic, since we cannot assign a truth value to it.

Simple statements joined together by **connectives** ("and", "or", "not", "if then", and "if and only if") result in compound statements. Note that compound statements can also be formed using "but", "however", or "never the less". A compound statement can be assigned a truth value.

Conditional statements are frequently written in "if-then" form. The "if" clause of the conditional is known as the **hypothesis**, and the "then" clause is called the **conclusion**. In a proof, the hypothesis is the information that is assumed to be true, while the conclusion is what is to be proven true. A conditional is considered to be of the form: **If p, then q** where p is the hypothesis and q is the conclusion.

$p \rightarrow q$ is read "if p then q".

~ (statement) is read "it is not true that (statement)".

The **converse** of a conditional requires the conclusion and hypothesis to trade places in the statement. For instance: the converse of the statement "If it is snowing outside, then it is cold" is "If it is cold outside, then it is snowing." Note by this example that not every converse created from a conditional will be true. If the converse is true, the conditional is said to be reversible.

The **inverse** of a conditional is the negation of both the hypothesis and conclusion. Again, given "If it is snowing outside, then it is cold," the inverse is "If it is not snowing outside, then it is not cold."

Quantifiers are words describing a quantity under discussion. These include words like "all", "none" (or "no"), and "some".

Negation of a Statement—If a statement is true, then its negation must be false (and vice versa).

A Summary of Negation Rules	
Statement	Negation
(1) q	(1) *not* q
(2) *not* q	(2) q
(3) π *and* s	(3) (not π) *or* (not s)
(4) π *or* s	(4) (not π) *and* (not s)
(5) if p, then q	(5) (p) *and* (not q)

Example: Select the statement that is the negation of "some winter nights are not cold".

    A. All winter nights are not cold.
    B. Some winter nights are cold.
    C. All winter nights are cold.
    D. None of the winter nights are cold.

The correct answer is D.

Negation of "some are" is "none are". So the negation statement is "none of the winter night is cold".

Example: Select the statement that is the negation of "if it rains, then the beach party will not be held".

    A. If it does not rain, then the beach party will be held.
    B. If the beach party is held, then it will not rain.
    C. It does not rain and the beach party will be held.
    D. It rains and the beach party will be held.

The correct answer is D.

Negation of "if p, then q" is "p and (not q)". So the negation of the given statement is "it rains and the beach party will be held".

Example: Select the negation of the statement "If they get elected, then all politicians go back on election promises".

    A. If they get elected, then many politicians go back on election promises.
    B. They get elected and some politicians go back on election promises.
    C. If they do not get elected, some politicians do not go back on election promises.
    D. None of the above statements is the negation of the given statement.

The correct answer is D.

Identify the key words of "if…then" and "all…go back". The negation of the given statement is "they get elected and none of the politicians go back on election promises". So select response D, since A, B, and C, statements are not the negations.

Example: Select the statement that is the negation of "the sun is shining bright *and* I feel great".

    A. If the sun is not shining bright. I do not feel great.
    B. The sun is not shining bright and I do not feel great.
    C. The sun is not shining bring or I do not feel great.
    D. the sun is shining bright and I do not feel great.

The correct answer is C.

The negation of "r and s" is "(not r) or (not s)". So the negation of the given statement is "the sun is *not* shining bright *or* I do not feel great".

Conditional statements can be diagrammed using a **Venn diagram**. A diagram can be drawn with one circle inside another circle. The inner circle represents the hypothesis. The outer circle represents the conclusion. If the hypothesis is taken to be true, then you are located inside the inner circle. If you are located in the inner circle then you are also inside the outer circle, so that proves the conclusion is true.

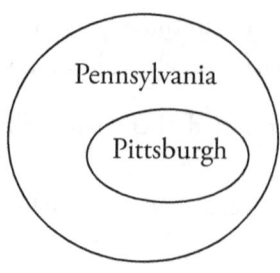

Example: If an angle has a measure of 90 degrees, then it is a right angle.

In this statement "an angle has a measure of 90 degrees" is the hypothesis. In this statement "it is a right angle" is the conclusion.

Example: If you are in Pittsburgh, then you are in Pennsylvania.

In this statement "you are in Pittsburgh" is the hypothesis. In this statement "you are in Pennsylvania" is the conclusion.

Deductive Reasoning and Validity

Deductive reasoning is the process of arriving at a conclusion based on other statements that are all known to be true.

A symbolic argument consists of a set of premises and a conclusion in the format of of if [Premise 1 and premise 2] then [conclusion].

An argument is **valid** when the conclusion follows necessarily from the premises. An argument is **invalid** or a fallacy when the conclusion does not follow from the premises.

There are 4 standard forms of valid arguments which must be remembered.

1. Law of Detachment   If p, then q   (premise 1)
   p,   (premise 2)
   Therefore, q
2. Law of Contraposition   If p, then q
   not q,
   Therefore not p
3. Law of Syllogism   If p, then q
   If q, then r
   Therefore if p, then r
4. Disjunctive Syllogism   p or q
   not p
   Therefore, q

Example: Can a conclusion be reached from these two statements?

A. All swimmers are athletes.
All athletes are scholars.

In "if-then" form, these would be:

> If you are a swimmer, then you are an athlete.
> If you are an athlete, then you are a scholar.

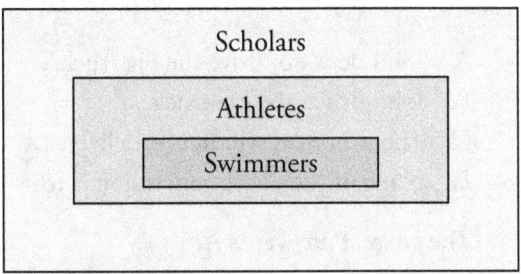

Clearly, if you are a swimmer, then you are also an athlete. This includes you in the group of scholars.

B. All swimmers are athletes.
All wrestlers are athletes.

In "if-then" form, these would be:

> If you are a swimmer, then you are an athlete.
> If you are a wrestler, then you are an athlete.

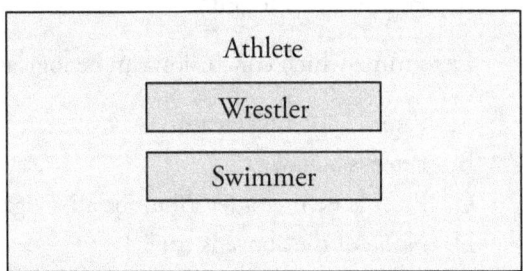

Clearly, if you are a swimmer or a wrestler, then you are also an athlete. This swimmer does not allow you to come to any other conclusions.

A swimmer may or may NOT also be a wrestler. Therefore, **no conclusion is possible**.

Suppose that these statements were given to you, and you are asked to try to reach a conclusion. The statements are:

Example: Determine whether statement A, B, C, or D can be deduced from the following:

(i) If John drives the big truck, then the shipment will be delivered.
(ii) The shipment will not be delivered.

A. John does not drive the big truck.
B. John drives the big truck.
C. The shipment will not be delivered.
D. None of the above conclusion is true.

The correct answer is A.

Let p: John drives the big truck.
 q: The shipment is delivered.

statement (i) gives $p \rightarrow q$, statement (ii) gives ~q. This is the Law of Contraposition. Therefore, the logical conclusion is ~p or "John does not drive the big truck".

Example: Given that:

(i) Peter is a Jet Pilot or Peter is a Navigator.
(ii) Peter is not a Jet Pilot

Determine which conclusion can be logically deduced.

A. Peter is not a Navigator.
B. Peter is a Navigator.
C. Peter is neither a Jet Pilot nor a Navigator.
D. None of the above is true.

The correct answer is B.

Let p: Peter is a Jet Pilot
 q: Peter is a Navigator.

So we have $p \vee q$ from statement (i)
 ~p from statement (ii)

Try These:

What conclusion, if any, can be reached? Assume each statement is true, regardless of any personal beliefs.

1. If the Red Sox win the World Series, I will die.
   I died.
2. If an angle's measure is between 0° and 90°, then the angle is acute.
   Angle B is not acute.
3. Students who do well in geometry will succeed in college.
   Annie is doing extremely well in geometry.
4. Left-handed people are witty and charming.
   You are left-handed.

Question #1   The Red Sox won the World Series.
Question #2   Angle B is not between 0 and 90 degrees.
Question #3   Annie will do well in college.
Question #4   You are witty and charming.

Set Theory

Consider a Venn diagram with a finite set of elements, such as letters.

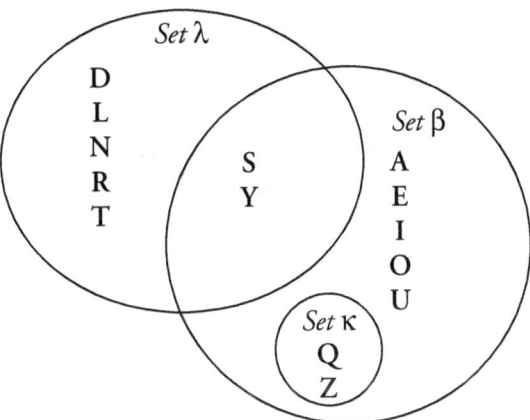

Various symbols and statements can be used to describe the sets and their content in a sort of mathematical shorthand.

The Venn diagram shows *Set* β, *Set* λ. and *Set* κ
- λ = {D, L, N, R, T, S, Y)
- β = {A, E, I, O, U, S, Y, Q, Z}
- κ = {Q, Z}

R is an **element** of Set λ, but not Set β.

- $R \in \lambda$
- $R \notin \beta$

The elements Y and S are members of both sets λ and β. This is usually described as the **intersection** of the two sets.

- $\lambda \cap \beta = \{S, Y\}$

Furthermore, as set κ is completely inside set β, it is classified as more than just an intersection. The set κ is a **subset** of β.

- $\kappa \subseteq \beta$

Two or more sets can be combined by joining them in a **union**.

- $\lambda \cup \kappa = \{D, L, N, R, T, S, Y, Q, Z\}$
- $\beta \cup \kappa = \beta$ (note: a union does not duplicate elements)

Another operation between two sets is the **Cartesian product**. The product "$\lambda \times \kappa$" (read "cross") is the set of all ordered pairs (b, c) where b is a member of λ and c is a member of κ.

- $\lambda \times \kappa = \{$ (D, Q), (L, Q), (N, Q), (R, Q), (T, Q), (S, Q), (Y, Q), (D, Z), (L, Z), (N, Z), (R, Z), (T, Z), (S, Z), (Y, Z) $\}$

Oftentimes it is impractical to list every element in a set. Some of these cases can be as follows:

- $\Omega = \{\Re\}$ denotes the set of all real numbers
- while the set $\varphi = \{x \in \Re \mid x < 2\}$ contains "all real numbers x *such that* x is less than 2."
- $\{n \mid n > 0\}$ describes a set of numbers greater than zero. Note that this set does not have a name, which can often occur.

Numbers

Order of Operations

When simplifying algebraic expressions we use the following order:
1. Perform operations within a parenthesis.
2. Evaluate exponents.
3. Multiply and divide from left to right.
4. Add and subtract from left to right.

Example: $3 + 2(4 + 3)^2 - 10 \div 5$ Perform operations in parentheses.
$= 3 + 2(7)^2 - 10 \div 5$
$= 3 + 2(49) - 10 \div 5$ Evaluate exponents.
$= 3 + 98 - 2$ Multiply and divide from left to right.
$= 99$ Add and subtract from left to right.

Properties of Operations

Properties of operations are rules that apply for addition, subtraction, multiplication, or division of real numbers.

Commutative Property

You can change the order of the terms or factors as follows.
For addition: $a + b = b + a$
For multiplication: $ab = ba$
This rule does not apply for division and subtraction.

Examples: $17 + 23 = 23 + 17 = 40$
$8 \times 19 = 19 \times 8 = 152$

Associative Property

You can regroup the terms as you like.
For addition: $a + (b + c) = (a + b) + c$
For multiplication: $a(bc) = (ab)c$
This rule also does not apply for division and subtraction.

Example: $(-2 + 7) + 5 = -2 + (7 + 5)$
$5 + 5 = -2 + 12 = 10$

Example: $(3 \times -7) \times 5 = 3 \times (-7 \times 5)$
$-21 \times 5 = 3 \times -35 = -105$

Identity Properties

Adding 0 to a number results in that number, with no change (additive identity of 0); multiplying a number by 1 results in that number, with no change (multiplicative identity of 1).

For addition: $a + 0 = a$ (additive identity of 0)
For multiplication: $a \times 1 = a$ (multiplicative identity of 1)

Example: $17 + 0 = 17$

Example: $-34 \times 1 = -34$

Inverse Properties

The additive inverse of a number a is the number that when added to a results in zero. The multiplicative inverse (or reciprocal) of a number a is the number that when multiplied by a results in 1.

For addition: $a + (-a) = 0$

Example: $25 + -25 = 0$

For multiplication: $a \times \dfrac{1}{a} = 1$

Example: $5 \times \dfrac{1}{5} = 1$

The additive inverse of a is $(-a)$. The reciprocal or multiplicative inverse of a is $\dfrac{1}{a}$. The sum of any number and its additive inverse is 0. The product of any number and its reciprocal is 1.

Distributive Property of Multiplication Over Addition and Subtraction

This property allows us to operate on terms within parentheses without first performing operations within the parentheses. This is especially helpful when terms within the parentheses cannot be combined.

$$a(b + c) = ab + ac$$

Example: $6 \times (-4 + 9) = (6 \times -4) + (6 \times 9)$
$6 \times 5 = -24 + 54 = 30$

To multiply a sum by a number, multiply each addend by the number, then add the products.

Summary of the Properties of Operations

Property	of Addition	of Multiplication
Commutative	$a + b = b + a$	$ab = ba$
Associative	$a + (b + c) = (a + b) + c$	$a(bc) = (ab)c$
Identity	$a + 0 = a$	$a \times 1 = a$
Inverse	$a + (-a) = 0$	$a \times \frac{1}{a} = 1, a \neq 0$
Distributive property of multiplication over addition and subtraction	$a(b + c) = ab + ac$	$a(b - c) = ab - ac$

Number Sets, Classifications, and Theory

Natural (Counting) Numbers

The set of **natural numbers**, $\mathbb{N}$, includes 1, 2, 3, 4, … (By some definitions, $\mathbb{N}$ includes zero.) The natural numbers are sometimes called the counting numbers (especially if the definition of $\mathbb{N}$ excludes zero).

The set $\mathbb{N}$ obeys the properties of associativity, commutativity, distributivity and identity for multiplication and addition (assuming, for the case of addition, that zero is included in some sense in the natural numbers). The set of natural numbers does not contain additive or multiplicative inverses, however, as there are no noninteger fractions or negative numbers.

Natural numbers can be either even or odd. **Even** numbers are evenly divisible by two; **odd** numbers are not evenly divisible by two (alternatively, they leave a remainder of one when divided by two).

Properties of Odd and Even Numbers

Addition

Rule	Odd + Odd = Even	Odd + Even = Odd	Even + Odd = Odd	Even + Even = Even
Example	5 + 7 = 12	9 + 22 = 31	16 + 13 = 29	28 + 4 = 32

Multiplication

Rule	Odd × Odd = Odd	Odd × Even = Even	Even × Odd = Even	Even × Even = Even
Example	5 × 7 = 35	9 × 22 = 198	16 × 13 = 208	28 ×4 = 112

Composite and Prime Numbers

Any natural number n that is divisible by at least one number that is not equal to 1 or n is called a **composite number**. A **prime number** is a natural number n that is divisible by two numbers only: 1 and n. The number 1 does not qualify as a prime number because it is divisible by only one number, itself, whereas a prime is divisible by exactly two numbers.

Example: 91 is divisible by 7 and 13 in addition to being divisible by 91 and 1, so 91 is a composite number. 93 is divisible only by 93 and 1, so 93 is a prime number.

Divisibility Tests for Natural Numbers

In many cases, it is possible to say whether a natural number is divisible by a certain factor without actually doing the division, by applying a divisibility test. There is such a test for every number from 1 to 12 except 7.

1. Every natural number is divisible by 1.
2. A number is divisible by 2 if that number is an even number (i.e., the last digit is 0, 2, 4, 6 or 8). Consider a number defined by the digits a, b, c and d (for instance, 1,234). Rewrite the number as follows. The value of the number is $1000a + 100b + 10c + d$ The first three terms are divisible by 2. Thus, the number is only divisible by 2 if d is divisible by two. For example, the last digit of 1,354 is 4, so it is

divisible by 2. On the other hand, the last digit of 240,685 is 5, so it is not divisible by 2.

3. A number is divisible by 3 if the sum of its digits is evenly divisible by 3. Consider a number defined by the digits a, b, c and d. The number can be written as

   $1000a + 100b + 10c + d$ The number can also be rewritten as
   $(999 + 1)a + (99 + 1)b + (9 + 1)c + d$
   $999a + 99b + 9c + (a + b + c + d)$

   Note that the first three terms in the above expression are all divisible by 3. Thus, the number is evenly divisible by 3 only if $a + b + c + d$ is divisible by 3. The same logic applies regardless of the size of the number. This proves the rules for divisibility by 3. The sum of the digits of 964 is $9 + 6 + 4 = 19$. Since 19 is not divisible by 3, neither is 964. The sum of the digits of 86,514 is $8 + 6 + 5 + 1 + 4 = 24$. Since 24 is divisible by 3, 86,514 is also divisible by 3.

4. A number is divisible by 4 if the last two digits of the number are evenly divisible by 4. Let a number $abcd$ be defined by the digits a, b, c and d.

   $1000a + 100b + 10c + d$

   The number can also be written as $100(10a + b) + 10c + d$

   Since 100 is divisible by 4, $100(10a + b)$ is also divisible by 4. Thus, $abcd$ is divisible by 4 only if the two-digit number cd is divisible by 4.

   The number 113,336 ends with the number 36 for the last two digits. Since 36 is divisible by 4, 113,336 is also divisible by 4. The number 135,626 ends with the number 26 for the last two digits. Since 26 is not evenly divisible by 4, 135,626 is also not divisible by 4.

5. A number is divisible by 5 if the number ends in either a 5 or a 0. Once again, imagine a number with the digits $abcd$. The number equals

   $1000a + 100b + 10c + d$

   The first three terms are evenly divisible by 5, but the last term is only evenly divisible by 5 if d is divisible by 5, that is, if the number ends in a 0 or a 5. For instance, 225 ends with a 5, so it is divisible by

5. The number 470 is also divisible by 5 because its last digit is a 0. The number 2,358 is not divisible by 5 because its last digit is an 8.
6. A number is divisible by 6 if the number is divisible by both 2 and 3. Thus, any even number that is divisible by 3 is also divisible by 6. For instance, 4,950 is an even number and its digits add up to $4 + 9 + 5 + 0 = 18$. Since it is even and the sum of its digits is divisible by 3, the number 4,950 is divisible by 3 and by 6 as well. On the other hand, 326 is an even number, but its digits add up to 11, which is not divisible by 3. Therefore, 326 is not divisible by 3 or by 6.
7. There is no easy test of divisibility by 7.
8. A number is divisible by 8 if the number in its last three digits is evenly divisible by 8. The logic for the proof of this case follows that of numbers divisible by 2 and 4. Taking a number with digits *abcd*, we can write its value as

    $1000a + 100b + 10c + d$

    The first term is divisible by 8, so the number as a whole is divisible by 8 only if the last three terms compose a number divisible by 8.

    Example: The number 113,336 ends with the 3-digit number 336. Since 336 is divisible by 8, then 113,336 is also divisible by 8. The number 465,628 ends with the 3-digit number 628. Since 628 is not evenly divisible by 8, then 465,628 is also not divisible by 8.

9. A number is divisible by 9 if the sum of its digits is evenly divisible by 9. The logic for the proof of this case follows that for the case of numbers that are divisible by 3 and 6. The sum of the digits of 874, for example, is $8 + 7 + 4 = 19$. Since 19 is not divisible by 9, neither is 874. The sum of the digits of 116,514 is $1 + 1 + 6 + 5 + 1 + 4 = 18$. Since 18 is divisible by 9, 116,514 is also divisible by 9.
10. A number is divisible by 10 if the last digit is zero.
11. A number is divisible by 11 if the sum of every other digit differs from the sum of the intervening digits by 0 or a multiple of 11.

    Example: The sum of every other digit of 132,847 is $1 + 2 + 4 = 7$. The sum of the intervening digits is $3 + 8 + 7 = 18$.

The difference of 18 and 7 is 11, so 132,847 is divisible by 11.

The sum of every other digit of 61,955 is $6 + 9 + 5 = 20$. The sum of the intervening digits is $1 + 5 = 6$. The difference of 20 and 6 is 14, so 61,955 is not divisible by 11.

12. A number is divisible by 12 if it is divisible by both 4 and 3: the last two digits are divisible by 4 and the sum of the digits is divisible by 3.

    Example: The last two digits of 864 are 64, a number divisible by 4. The sum of the digits is $8 + 6 + 4 = 18$, which is divisible by 3. Therefore, 864 is divisible by 12.

The strategies used in these divisibility tests can be extended to other higher factors. A number is divisible by 15, for instance, if it is divisible by 3 and 5, and it is divisible by 44 if it is divisible by 4 and 11.

Another concept closely related to divisibility involves analyzing numbers by their factors. For instance, 3 and 5 are factors of 15. GCF is the abbreviation for the **greatest common factor**. The GCF is the largest number that is a factor of all the numbers given in a problem. The GCF can be no larger than the smallest number given in the problem. If no other number is a common factor, then the GCF will be the number 1. To find the GCF, list all possible factors of the smallest number given (include the number itself). Starting with the largest factor (which is the number itself), determine if it is also a factor of all the other given numbers. If so, that is the GCF. If that factor does not work, try the same method on the next smaller factor. Continue until a common factor is found. This is the GCF. Note: There can be other common factors besides the GCF.

Example: Find the GCF of 12, 20, and 36.

The smallest number in the problem is 12. The factors of 12 are 1, 2, 3, 4, 6 and 12. 12 is the largest factor, but it does not divide evenly into 20. Neither does 6, but 4 will divide into both 20 and 36 evenly. Therefore, 4 is the GCF.

Example: Find the GCF of 14 and 15.

Factors of 14 are 1, 2, 7 and 14. 14 is the largest factor, but it does not divide evenly into 15. Neither does 7 or 2. Therefore, the only factor common to both 14 and 15 is the number 1, which is the GCF.

The Fundamental Theorem of Arithmetic

Every integer greater than 1 can be written uniquely in the form

$$p_1^{e1} p_2^{e2} \cdots p_k^{ek}$$

The pi are distinct prime numbers and the ei are positive integers. Any integer $n > 1$ that is divisible by at least one positive integer that is not equal to one or n is called a **composite number**. A natural number n that is only divisible by one and n is called a **prime number**. Expressing an integer by the product of its prime factors is called prime factorization.

Example: Find the prime factorization of 68.

Create a factor tree. Keep dividing each branch until a prime number is reached.

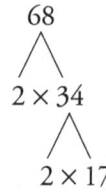

This technique shows that $68 = 2 \times 2 \times 17$ or $2^2 \times 17$

LCM is the abbreviation for **least common multiple**. The least common multiple of a group of numbers is the smallest number that all of the given numbers will divide into. The least common multiple will always be the largest of the given numbers or a multiple of the largest number.

Example: Find the LCM of 20, 30 and 40.

The largest number given is 40, but 30 will not divide evenly into 40. The next multiple of 40 is 80 (2 × 40), but 30 will not divide evenly into 80 either. The next multiple of 40 is 120. 120

is divisible by both 20 and 30, so 120 is the LCM (least common multiple).

Example: Find the LCM of 96, 16 and 24.

The largest number is 96. 96 is divisible by both 16 and 24, so 96 is the LCM.

The fundamental theorem of arithmetic can be used to show that **every fraction is equivalent to a unique fraction where the numerator and denominator are relatively prime.**

Given a fraction $\frac{a}{b}$, the integers a and b can both be written uniquely as a product of prime factors.

$$\frac{a}{b} = \frac{p_1^{x_1} p_2^{x_2} p_3^{x_3} \cdots p_n^{x_n}}{q_1^{y_1} q_2^{y_2} q_3^{y_3} \cdots q_m^{y_m}}$$

When all the common factors are cancelled, the resulting numerator a_1 (the product of remaining factors $p_n^{x_n}$) and the resulting denominator b_1 (the product of remaining factors $q_m^{y_m}$) have no common divisor other than 1; i.e., they are **relatively prime**. Since, according to the Fundamental Theorem of Arithmetic, the initial prime decomposition of the integers a and b is unique, the new reduced fraction $\frac{a_1}{b_1}$ is also **unique**. Hence, any fraction is equivalent to a unique fraction where the numerator and denominator are relatively prime.

Real Numbers

The following chart shows the relationships among the subsets of the real numbers.

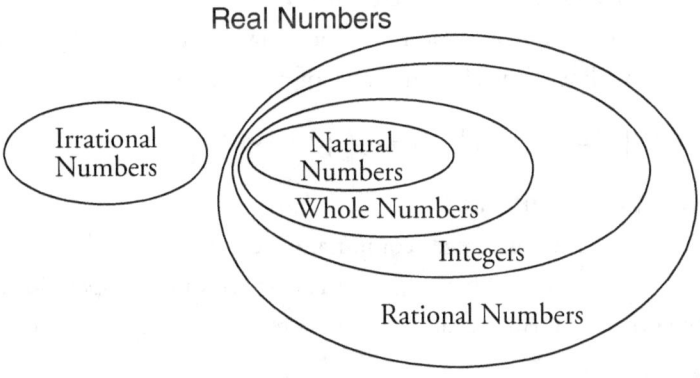

Real numbers are denoted by ℝ and are numbers that can be shown by an infinite decimal representation such as 3.286275347. . . . Real numbers include **rational numbers**, numbers that can be represented as the quotient of two integers, such as 242 and $-\frac{23}{129}$, and **irrational numbers**, such as $\sqrt{2}$ and π, which cannot. All real numbers can be represented as points along an infinite number line.

Real numbers are to be distinguished from imaginary numbers.

Real numbers are classified as follows:

Natural Numbers Denoted by ℕ	The counting numbers. 1, 2, 3, . . .
Whole Numbers	The counting numbers along with zero. 0, 1, 2, 3, . . .
Integers Denoted by ℤ	The counting numbers, their negatives, and zero. . . . , −2, −1, 0, 1, 2, . . .
Rationals, Denoted by ℚ	All of the fractions that can be formed using whole numbers. Zero cannot be the denominator. In decimal form, these numbers will be either terminating or repeating decimals. Simplify square roots to determine if the number can be written as a fraction.
Irrationals	Real numbers that cannot be written as a fraction. The decimal forms of these numbers neither terminate nor repeat. Examples include π, e, and $\sqrt{2}$.

Calculating within Number Systems

The absolute value of a number is the distance of the number from zero on the number line. "The absolute value of n" is written as $|n|$.

Examples: $|-122| = 122, |8.7| = 8.7, |3 - 10| = |-7| = 7$

Operations with Integers

While the basics of integer computation can be reasoned on a number line, it is essential to quickly recognize the patterns of basic operations.

The **addition** of two negative numbers results in a negative number.

Example: $-3 + -8 = -11$

When adding a positive and a negative number, the number with the greater absolute value dominates the sign of the answer. The value is the difference between the absolute values.

Example: $-6 + 14 = 8$ (positive answer, since $14 > 6$)

Example: $-9 + 5 = -4$ (negative answer, since $9 > 5$)

When performing **subtraction** on integers, the operation can be written as addition of the opposite.

Example: $7 - (-5)$ is rewritten as $7 + 5 = 12$

When needing to determine the sign of an answer in a **multiplication** or **division** problem, simply count up the number of negatives. An even number of negatives will result in a positive product or quotient, an odd number will be negative.

Example: $\frac{-8(-5)}{-4} = -10$ (negative answer from an odd number of negatives)

Operations with Rational Numbers

Division of fractions should be rewritten as multiplication of the reciprocal.

Example: $\frac{3}{7} \div \frac{1}{3} = \frac{3}{7} \times \frac{3}{1} = \frac{9}{7}$

Rational number **addition** relies on the distributive property of multiplication over addition and the understanding that multiplication by any number by one yields the same number.

Example: Consider the addition of 1/4 to 1/3 by means of common denominator.

$$\frac{1}{4} + \frac{1}{3} = \frac{3}{3}\left(\frac{1}{4}\right) + \frac{4}{4}\left(\frac{1}{3}\right) = \left(\frac{3}{12}\right) + \left(\frac{4}{12}\right) = \frac{7}{12}$$

Recognize that $\frac{3}{3}$ and $\frac{4}{4}$ both equal unity, or 1.

Operations with Irrational Numbers

Note: Radicals ($\sqrt{}$) are inverse operators of exponents and are represented in the form $\sqrt[n]{a^x}$, where

- n is called the index or root (assumed to be 2 if omitted)
- a^x is called the radicand
- x is the exponent or power of 'a'

In the notation above, we are finding the nth root of a^x. In other words, we want to find the number that, when multiplied by itself n times, yields a^x. An example is shown below for the number 16.

$$\sqrt{16} = \sqrt[2]{16}$$
$$\sqrt{16} = \pm 4$$

Both the positive and negative values of 4 are solutions because the square of +4 and the square of −4 are both 16. The number +4 is called the principal square root of 16, because the principal square root is the only one that makes sense in some cases (for example, in measurements).

A radical should be simplified whenever possible. This means "answering" all perfect powers that are factors of the radicand. Or, in other words, noting that when the radicand is raised to a power that is equal to the index, the root operation will cancel out the power operation.

Example: Perform the indicated operations.

a) $\sqrt[3]{2^3}$

The power and the index are equal to each other. Thus, $\sqrt[3]{2^3} = 2$

b) $2\sqrt{32}$

Rewrite the number 32 as follows:

$$32 = 2 \times 16 = 2 \times 4 \times 4 = 2 \times 4^2$$

Thus, $2\sqrt{32} = 2\sqrt{2 \times 4^2} = 2 \times 4\sqrt[2]{2}$ or $8\sqrt{2}$

The example above shows that radicals can be separated (or joined) by multiplication. This is not the case for addition or subtraction. That is $\sqrt{7} \neq \sqrt{5} + \sqrt{2}$. We can only add or subtract radicals that have the same index and the *same radicand.*

Example: $2\sqrt{5} + 3\sqrt{5} = 5\sqrt{5}$

Example: $5\sqrt[3]{2} - 3\sqrt[3]{2} = 2\sqrt[3]{2}$

If the radicand is raised to a power different from the index, convert the radical to its exponential form and apply laws of exponents.

Example: $\sqrt[6]{9} = \sqrt[6]{3^2} = 3^{2/6} = 3^{1/3} = \sqrt[3]{3}$

Applications of Numbers

Ratios

A **ratio** is a comparison of two numbers for the purpose of relating relative magnitudes. For instance, if a class had 11 boys and 14 girls, the ratio of boys to girls could be written one of three ways:

$$11 : 14 \text{ or } 11 \text{ to } 14 \text{ or } \frac{11}{14}$$

The ratio of girls to boys is: $14 : 11$ or 14 to 11 or $\frac{14}{11}$

Ratios, like fractions, can be reduced if their terms have common factors. A ratio of 12 cats to 18 dogs, for example, reduces to $2 : 3$, 2 to 3 or $\frac{2}{3}$

Proportions

A **proportion** is an equation in which a fraction is set equal to another fraction. To solve the proportion, multiply each numerator times the other fraction's denominator. Set these two products equal to each other and solve the resulting equation. This is called cross-multiplying the proportion.

Example: Find x given the proportion $\frac{4}{15} = \frac{x}{60}$

To solve for x, cross-multiply.

$$(4)(60) = (15)(x)$$
$$240 = 15x$$
$$16 = x$$

Example: Find x given the proportion $\frac{x+3}{3x+4} = \frac{2}{5}$

To solve for x, cross-multiply.

$$5(x+3) = 2(3x+4)$$
$$5x + 15 = 6x + 8$$
$$7 = x$$

The mathematics of solving for variables in proportions is not difficult. The key to solving a problem involving proportions is constructing the proportion correctly. As noted above, this requires carefully reading the problem, followed by careful identification of the related values and construction of the appropriate ratios.

Proportions can be used to solve word problems whenever relationships are compared. Some situations include scale drawings and maps, similar polygons, speed, time and distance, cost, and comparison shopping.

Example: Which is the better buy, 6 items for $1.29 or 8 items for $1.69?

Find the unit price.

$$\frac{6}{1.29} = \frac{1}{x} \qquad \frac{8}{1.69} = \frac{1}{x}$$
$$6x = 1.29 \qquad 8x = 1.69$$
$$x = 0.215 \qquad x = 0.21125$$

Thus, 8 items for $1.69 is the better buy.

Example: A car travels 125 miles in 2.5 hours. How far will it go in 6 hours?

Write a proportion comparing distance and time.

$$\frac{125}{2.5} = \frac{x}{6}$$
$$\frac{miles}{hours} : \qquad 2.5x = 750$$
$$x = 300$$

The car will travel 300 miles in 6 hours.

Example: The scale on a map is $\frac{3}{4}$ inch = 6 miles. What is the actual distance between two cities if they are $1\frac{1}{2}$ inches apart on the map? Write a proportion comparing the scale to the actual distance.

$$\begin{array}{cc} \text{scale} & \text{actual} \end{array}$$

$$\frac{\frac{3}{4}}{1\frac{1}{2}} = \frac{6}{x}$$

$$\frac{3}{4}x = 1\frac{1}{2} \times 6$$

$$\frac{3}{4}x = 9$$

$$x = 12$$

Thus, the actual distance between the cities is 12 miles.

Measurement

Measurement is a way of quantifying characteristics of physical matter, time, size, and space. Each characteristic has its own basis for measurement. Measuring length, for instance, is distinctly separate from measuring time. When taking measurements, the precision depends on the circumstances and relative size. The distance between two towns does not need to be measured to the nearest thousandth of a mile. The nearest whole mile or tenth of a mile is usually sufficient. Additionally, the measurement tool being used influences the amount of precision expected. Most rulers measure to the nearest millimeter and bathroom scales to the nearest pound or tenth of a pound. Therefore, if any calculations need to be performed with measurements, consider such precision when presenting the answer. For instance, a velocity calculated from data in whole feet and seconds should be rounded to the nearest whole or tenth of a number. Any further decimal places would be unreliable.

The following units of measure are fairly commonplace:

Measurements of Length (English System)

12 inches (in)	=	1 foot (ft)
3 feet (ft)	=	1 yard (yd)
1760 yards (yd)	=	1 mile (mi)

Measurements of Length (Metric System)

kilometer (km)	=	1000 meters (m)
hectometer (hm)	=	100 meters (m)

decameter (dam)	=	10 meters (m)
meter (m)	=	1 meter (m)
decimeter (dm)	=	1/10 meter (m)
centimeter (cm)	=	1/100 meter (m)
millimeter (mm)	=	1/1000 meter (m)

Conversion of Length from English to Metric

1 inch	=	2.54 centimeters
1 foot	≈	30 centimeters
1 yard	≈	0.9 meters
1 mile	≈	1.6 kilometers

Measurements of Weight (English System)

28 grams (g)	=	1 ounce (oz)
16 ounces (oz)	=	1 pound (lb)
2000 pounds (lb)	=	1 ton (t) (short ton)
1.1 ton (t)	=	1 ton (t)

Measurements of Weight (Metric System)

kilogram (kg)	=	1000 grams (g)
gram (g)	=	1 gram (g)
milligram (mg)	=	1/1000 gram (g)

Conversion of Weight from English to Metric

1 ounce	≈	28 grams
1 pound	≈	0.45 kilogram
	≈	454 grams

Measurement of Volume (English System)

8 fluid ounces (oz)	=	1 cup (c)
2 cups (c)	=	1 pint (pt)
2 pints (pt)	=	1 quart (qt)
4 quarts (qt)	=	1 gallon (gal)

Measurement of Volume (Metric System)

kiloliter (kl)	=	1000 liters (l)
liter (l)	=	1 liter (l)
milliliter (ml)	=	1/1000 liters (ml)

Conversion of Volume from English to Metric

1 teaspoon (tsp)	≈	5 milliliters
1 fluid ounce	≈	15 milliliters
1 cup	≈	0.24 liters
1 pint	≈	0.47 liters
1 quart	≈	0.95 liters
1 gallon	≈	3.8 liters

Measurement of Time

1 second	=	
1 minute	=	60 seconds
1 hour	=	60 minutes
1 day	=	24 hours
1 week	=	7 days
1 year	=	365 days
1 century	=	100 years

Note: (') represents feet and (") represents inches.

Often, the need arises to convert between units of measure. When calculating such changes, use the "unit cancelation method" to make appropriate multiplication or division choices. For instance, since there are 36 inches in a yard, the conversion value can be expressed as $\frac{36 \text{ in}}{1 \text{ yd}}$ or $\frac{1 \text{ yd}}{36 \text{ in}}$ as needed. Suppose the length of 48 inches needs to be converted into yards. Then inches need to be canceled out while yards need to be part of the answer.

Therefore consider: $\frac{48 \text{ in}}{1} \times \frac{1 \text{ yd}}{36 \text{ in}} = \frac{48}{36}\left(\frac{\text{in} \times \text{yd}}{\text{in}}\right) = \frac{4}{3}(\text{yd}) = 1\frac{1}{3}\text{yd}$

The unit cancelation method can be used to convert any measured property, and between all systems of measurement.

Length

Example: A car skidded 170 yards on an icy road before coming to a stop. How long is the skid distance in kilometers?

Since 1 yard ≈ 0.9 meters, multiply 170 yards by 0.9.

$$170 \text{ yd} \times 0.9 \frac{\text{m}}{\text{yd}} = 153 \text{ m}$$

Since 1000 meters = 1 kilometer, divide 153 by 1000.

$$\frac{153}{1000} = 0.153 \text{ kilometers}$$

Example: The distance around a race course is exactly 1 mile, 17 feet, and $9\frac{1}{4}$ inches. Approximate this distance to the nearest tenth of a foot.

Convert the distance to feet.

$$1 \text{ mile} = 1760 \text{ yards} = 1760 \text{ yd} \times \frac{3 \text{ ft}}{1 \text{ yd}} = 5280 \text{ feet.}$$

$$9\frac{1}{4} \text{ inches} = \frac{37 \text{ in}}{4} \times \frac{1 \text{ ft}}{12 \text{ in}} = \frac{37 \text{ ft}}{48} \approx 0.77083 \text{ feet}$$

So 1 mile, 17 feet and $9\frac{1}{4}$ inches = 5280 + 17 + 0.77083 feet = 5297.7̲7083 feet.

Now, we need to round to the nearest tenth digit. (Note that this has been saved until the last step. That is, the only value to be rounded is that of the final answer.) The underlined 7 is in the tenth place. The digit in the hundredth place, also a 7, is greater than 5, the 7 in the tenths place needs to be rounded up to 8 to get a final answer of 5297.8 feet.

Weight

Example: Zachary weighs 150 pounds. Tom weighs 153 pounds. What is the difference in their weights in grams?

153 pounds − 150 pounds = 3 pounds

1 pound = 454 grams

$$3 \text{ lb} \left(\frac{454 \text{ g}}{1 \text{ lb}} \right) = 1362 \text{ grams}$$

Capacity

Example: Students in a fourth grade class want to fill a 3 gallon jug using cups of water. How many cups of water are needed?

1 gallon = 16 cups of water

3 gallons × 16 cups = 48 cups of water are needed.

Time

Example: It takes Cynthia 45 minutes to get ready each morning. How many hours does she spend getting ready each week?

$$45 \text{ minutes} \times 7 \text{ days} = 315 \text{ minutes}$$

$$\frac{315 \text{ minutes}}{60 \text{ minutes / hour}} = 5.25 \text{ hours}$$

Scientific notation is a more convenient method for writing very large and very small numbers. It employs two factors. The first factor is a number between 1 and 10. The second factor is a power of 10. This notation is a "shorthand" for expressing large numbers (like the weight of 100 elephants) or small numbers (like the weight of an atom in pounds).

Recall that:
$10^n = (10)^n$ Ten multiplied by itself n times.
$10^0 = 1$ Any nonzero number raised to power of zero is 1.
$10^1 = 10$
$10^2 = 10 \times 10 = 100$
$10^3 = 10 \times 10 \times 10 = 1000$ (kilo)
$10^{-1} = 1/10$ (deci)
$10^{-2} = 1/100$ (centi)
$10^{-3} = 1/1000$ (milli)
$10^{-6} = 1/1,000,000$ (micro)

Procedure to convert a large number to scientific notation: 46,368,000
1. Introduce a decimal point and decimal places. 46,368,000 = 46,368,000.0000
2. Make a mark between the two digits that give a number between −9.9 and 9.9. 4 ∧ 6,368,000.0000
3. Count the number of digit places between the decimal point and the ∧ mark. This number is the 'n'—the power of ten. So, 46,368,000 = 4.6368×10^7

Procedure to convert a small number to scientific notation: 0.00397
1. Decimal point is already in place.
2. Make a mark between 3 and 9 to get a one number between −9.9 and 9.9.
3. Move decimal place to the mark (3 hops).
 0.003 ∧ 97
 Motion is to the right, so n of 10^n is negative.
 Therefore, $0.00397 = 3.97 \times 10^{-3}$.

When calculating with numbers in scientific notation, simply follow the rules of exponents.

Example: $(2.3 \times 10^8)(2 \times 10^{-3}) = 4.6 \times 10^5$ (multiply like bases, add exponents)

Example: $(8 \times 10^4) \div (2 \times 10^{12}) = 4 \times 10^{-8}$ (divide like bases, subtract exponents)

Example: $(3 \times 10^4)^5 = 243 \times 10^{20} = 2.43 \times 10^{22}$ (raise a power to a power, multiply exponents)

Sample Test 1

Sample Test Questions

Directions: Read each item and select the best response.

1. Which of the following is closed under division?

    I. $\left\{\frac{1}{3}, 1, 3\right\}$

    II. $\{-1, 1\}$

    III. $\{-1, 0, 1\}$

    [A] I only

    [B] II only

    [C] III only

    [D] I and II

    [E] II and III

2. Which of the following is always composite if x is an odd positive integer and y is an even positive integer greater than 1?

    [A] $x + y$

    [B] $|x + y|$

    [C] $x + 2y$

    [D] $3x + y$

    [E] $3xy$

3. Find the LCM of 25, 18, and 24.

   [A] 1,200

   [B] 1,800

   [C] 2,400

   [D] 3,600

   [E] 10,800

4. Solve for x: $|3x| + 6 = 21$

   [A] [9, −5]

   [B] [−9, 5]

   [C] [−5, 0, 5]

   [D] [−5, 5]

   [E] [−9, 9]

5. Which graph represents the solution set for $x^2 - 5x > -6$?

   [A]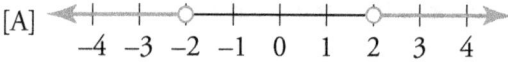

   [B]

   [C]

   [D]

   [E]

6. What is the equation of the graph shown below?

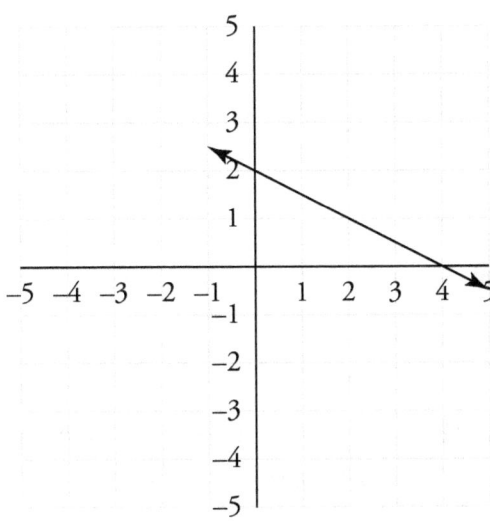

[A] $x + 2y = 4$

[B] $x - 2y = 4$

[C] $2x + y = 4$

[D] $x + 2y = -4$

[E] $x - 2y = -4$

7. Solve the following inequality: $-2x > 4$

[A] $x > -2$

[B] $x < -2$

[C] $x > 2$

[D] $x > -8$

[E] $x < 2$

8. Which equation represents a circle centered on the origin with radius 3?

   [A] $x^2 + y^2 = 3$

   [B] $x^2 + y^2 = 6$

   [C] $x^2 + y^2 = 9$

   [D] $x^2 + y^2 = 36$

   [E] $x^2 - y^2 = 9$

9. Given that D is a distance, M is a mass, T is a time, and V is a velocity, which of the following units could be used to measure $\frac{MTV}{D}$?

   [A] feet

   [B] meters

   [C] grams

   [D] seconds

   [E] miles per hour

10. Cubic meters are used to measure which of the following?

    [A] Distance

    [B] Length

    [C] Area

    [D] Volume

    [E] Mass

11. What figure best describes a data set in which many items are clustered near the median value with a smaller number of values less than or greater than the median at greater distances on each side?

    [A] A parabola

    [B] A normal curve

    [C] A line of best fit

    [D] A Cartesian curve

    [E] A Newtonian curve

12. If you prove a theorem by showing that an attempt to prove the opposite of the theorem leads to a contradiction, you are using the logical strategy called:

    [A] Inductive reasoning

    [B] Exhaustive proof

    [C] Proof by attraction

    [D] Direct proof

    [E] Indirect proof

13. Compute the area of the shaded region, given a radius of 7 meters. Point O is the center.

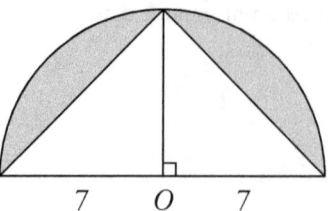

[A] 14.0

[B] 28.0

[C] 55.9

[D] 104.9

[E] 153.9

14. A garden measures 25 m by 40 m, including a circular fishpond with radius 3 m. What is the area of the garden not including the fishpond?

    [A] 101.7 m^2

    [B] 111.2 m^2

    [C] 971.7 m^2

    [D] 981.2 m^2

    [E] 990.6 m^2

15. The base of cone *A* has 3 times as great an area as the base of cone *B* but the height of cone *A* is only $\frac{1}{3}$ the height of cone *B*. Which statement is true?

   [A] Cone *A* has 9 times the volume of cone *B*

   [B] Cone *A* has 3 times the volume of cone *B*

   [C] Cone *A* and cone *B* have the same volume.

   [D] Cone *B* has 3 times the volume of cone *A*

   [E] Cone *B* has 9 times the volume of cone *A*

16. Find the area of the figure depicted below.

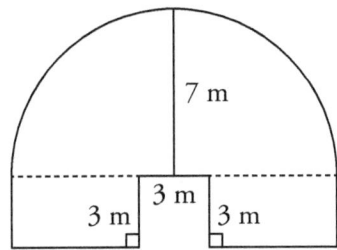

   [A] 109.9 m²

   [B] 118.9 m²

   [C] 142.9 m²

   [D] 144.9 m²

   [E] 186.9 m²

17. State the domain of the function $f(x) = \dfrac{2x-14}{x^2-9}$.

   [A] $x \neq 3$

   [B] $x \neq 3, 7$

   [C] $x \neq 3, -3$

   [D] $x \neq 7$

   [E] $x = 3, -3, 7$

18. Which of the following is a factor of the expression $6x^2 - 5x - 14$?

   [A] $3x + 7$

   [B] $6x + 7$

   [C] $6x - 7$

   [D] $6x - 5$

   [E] $x + 2$

19. Solve for x by factoring: $x^2 + x - 6 = 0$

   [A] $x = (-3, 2)$

   [B] $x = (3, -2)$

   [C] $x = (-6, 1)$

   [D] $x = (6, -1)$

   [E] no real solutions

20. Which of the following is equivalent to $\sqrt[b]{x^a}$?

   [A] $x^{\frac{a}{b}}$

   [B] $x^{\frac{b}{a}}$

   [C] $a^{\frac{x}{b}}$

   [D] $b^{\frac{x}{a}}$

   [E] $a^{\frac{b}{x}}$

21. Given $f(x) = 2x + 1$ and $g(x) = x^2 - 1$, determine $g(f(x))$.

   [A] $4x^2 + 4x - 1$

   [B] $4x^2 + 4x + 1$

   [C] $4x^2$

   [D] $4x^2 - 1$

   [E] $4x^2 + 4x$

22. Compute the median for the following data set: {9, 11, 18, 13, 12, 21}

   [A] 12

   [B] 12.5

   [C] 13

   [D] 14

   [E] 15.5

23. Which graph represents the equation $y = x^2 + 3x$?

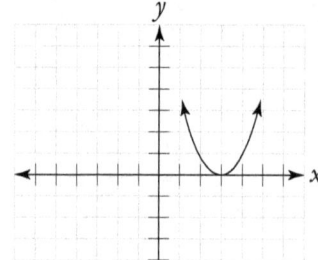

[A]

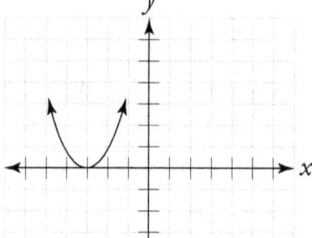

[B]

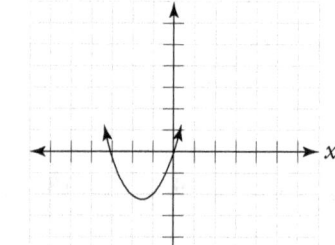

[C]

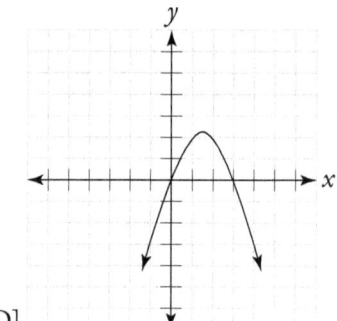

[D]

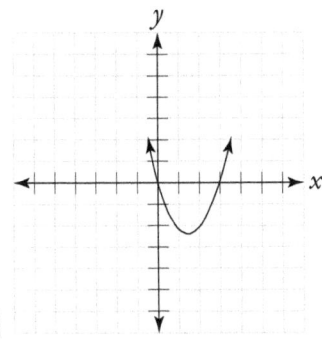

[E]

24. What would be the best measure of central tendency for the following collection of high temperatures on 10 successive days? {27, 24, 33, 24, 36, 65, 34, 30, 28, 29}

   [A] Mean

   [B] Either mean or median

   [C] Median

   [D] Mode

   [E] Either median or mode

25. If the correlation between two variables is zero, the association between the two variables is

   [A] Negative linear

   [B] Positive linear

   [C] Quadratic

   [D] Direct variation

   [E] Random

26. Which of the following is not a valid method of collecting statistical data?

   [A] Random sampling

   [B] Systematic sampling

   [C] Volunteer response

   [D] Weighted sampling

   [E] Cylindrical sampling

27. A jar contains 3 red marbles and 7 green ones. What is the probability that a marble picked at random from the jar will be red?

   [A] $\dfrac{1}{3}$

   [B] $\dfrac{1}{7}$

   [C] $\dfrac{3}{7}$

   [D] $\dfrac{3}{10}$

   [E] $\dfrac{7}{10}$

28. A die is rolled several times. What is the probability that a 6 will not appear before the fourth roll of the die?

   [A] $\dfrac{125}{216}$

   [B] $\dfrac{625}{1296}$

   [C] $\dfrac{1}{2}$

   [D] $\dfrac{5}{6}$

   [E] $\dfrac{1}{216}$

29. There is a 30% chance of rain this Saturday and a 30% chance of rain on Sunday as well. What is the chance of rain on both days?

[A] 9%

[B] 30%

[C] 49%

[D] 60%

[E] 70%

30. Which equation matches the data in the table?

x	3	4	5	6
y	7	8	9	10

[A] $y = 2x - 1$

[B] $y = 2x + 1$

[C] $y = -x + 10$

[D] $y = x + 4$

[E] $y = x - 4$

31. Which table could be generated by the equation: $y = x^2 + 2x - 1$?

[A]
x	1	2	3	4
y	2	5	8	11

[B]
x	1	2	3	4
y	4	9	16	25

[C]
x	1	2	3	4
y	1	5	11	19

[D]
x	1	2	3	4
y	2	7	13	21

[E]
x	1	2	3	4
y	2	7	14	23

32. The fees charged by a parking garage are as follows:

Hours	1	2	3	4	55
Fee	$12	$19	$26	$33	$40

How would you summarize the fees charged?

[A] $12 an hour

[B] $5 plus $7 per hour

[C] $15 an hour with a $3 discount

[D] $4 plus $8 per hour

[E] $3 plus $9 per hour

33. Which of the following is a solution to $x^2 + 4x + 4 = 25$?

   [A] 2

   [B] –2

   [C] –7

   [D] –3

   [E] 5

34. Solve the following system of equations:

$$2x + y = 8$$
$$4x + 2y = 20$$

   [A] $x = 2, y = 4$

   [B] $x = 3, y = 1$

   [C] $x = 4, y = 0$

   [D] no solutions

   [E] an infinite number of solutions

35. If an initial deposit of $10,000 is made to a savings account with interest compounded continuously at an annual rate of 6% how much money is in the account after 5 years?

   [A] $13,498.59

   [B] $3498.59

   [C] $13,382.26

   [D] $3,382.26

   [E] $13,000.00

36. A dance team comes prepared with a tango, a waltz, a disco number, a salsa routine, and a ballet selection. In how many different orders can they present their routines?

    [A] 5

    [B] 25

    [C] 120

    [D] 625

    [E] 3125

37. You can choose 3 selections from a buffet table with 8 dishes. How many different plates can you choose?

    [A] 6

    [B] 24

    [C] 56

    [D] 336

    [E] 6561

38. Leah has 4 blouses, 3 skirts, and 6 pairs of shoes. How many different outfits can she dress herself in?

    [A] 12

    [B] 13

    [C] 24

    [D] 72

    [E] 720

39. Hiroshi surveys his classmates to find what percent of them come to school on the bus, by car, by subway, by bicycle, or on foot. What is the best way to display his results?

[A] A line graph

[B] A box plot

[C] A stem-and-leaf plot

[D] A scatterplot

[E] A circle graph

40. Which equation could be used as a line of best fit for the scatterplot below?

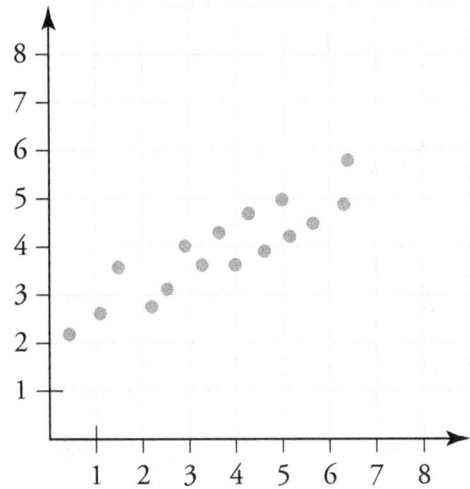

[A] $y = \dfrac{1}{2}x + 2$

[B] $y = 2x + 2$

[C] $y = -2x + 2$

[D] $y = \dfrac{1}{2}x - 2$

[E] $y = \dfrac{1}{2}x + 2$

41. To find the standard variation of a data set, you first compute the square of the distance of each data item from the mean of all the data items. Then what do you do?

   [A] Add all the squared distances and take the square root of the result.

   [B] Find the mean of the squared distances and take the square root of the result.

   [C] Multiply the squared distances and take the nth root of the result.

   [D] Multiply the square root of the sum of the squared distances by the mean of the squared distances.

   [E] Multiply the sum of the squared distances by the square root of the mean of the squared distances.

42. In which data set is the mode greater than the median?

   [A] {9,11,11,12,14}

   [B] {13,15,17,19,21]

   [C] {8,11,12,12,19}

   [D] {9,9,9,14,20}

   [E] {7,11,13,14,14}

43. Of the 200 students in the junior class, 8% are in the Spanish Club. How many juniors are in the Spanish Club?

   [A] 4

   [B] 8

   [C] 16

   [D] 20

   [E] 25

44. When Olga bought a boat for $1750, she paid an excise tax of $78.75. What was the percent of the tax?

 [A] 4.5%

 [B] 5.5%

 [C] 6.3%

 [D] 7%

 [E] 7.5%

45. A bank account pays 5% interest yearly. How large an amount would have to be deposited to earn $75 interest in a year?

 [A] $375

 [B] $875

 [C] $1200

 [D] $1500

 [E] $3750

46. A stock previously trading at $96 a share is now trading at $88 a share. What is the percent of change in the value of the stock?

 [A] −8%

 [B] −8.3%

 [C] −12%

 [D] −12.5%

 [E] −16%

47. The admission price to tour the Haunted House has been changed from $25 to $30 What is the percent of change in the admission price?

   [A] 5%

   [B] 16.7%

   [C] 20%

   [D] 25%

   [E] 30%

48. Eileen's Bakery had expenses of $62,500 last year and sales of $68,750. What was the profit as a percent of the expenses?

   [A] 6.25%

   [B] 10%

   [C] 12%

   [D] 15%

   [E] 16.7%

49. Tim's Typewriters had expenses of $26,200 last year and sales of $19,912. What was the loss as a percent of the expenses?

   [A] 7%

   [B] 8%

   [C] 16.7%

   [D] 20%

   [E] 24%

50. A stock that had been selling at $30 a share increased its share price by 20%. Later in the day the same stock suffered a 20% decrease in its share price. What was the price at the end of the day?

   [A] $24

   [B] $28.80

   [C] $30

   [D] $33

   [E] $36

51. A sweater is marked "25% off." The sale price is $36. What was the price before the discount?

   [A] $27

   [B] $32

   [C] $40

   [D] $45

   [E] $48

52. The sum of $1440 is deposited in a bank which pays 6% simple interest per year. After how many years will there be $1872 in the account?

   [A] 2.5 years

   [B] 3 years

   [C] 4 years

   [D] 5 years

   [E] 8 years

53. A bank pays 5% interest on deposits, compounded yearly. If $14,000 is deposited, how much will be in the account 3 years later?

   [A] $14,350

   [B] $15,435

   [C] $16,100

   [D] $16,206.75

   [E] $17,500

54. Which statement is logically equivalent to the following: If it's raining, my roof is leaking.

   [A] If my roof isn't leaking, it isn't raining.

   [B] If my roof is leaking, it's raining.

   [C] If it isn't raining, my roof isn't leaking.

   [D] If my roof is leaking, it's not raining

   [E] If it's raining, my roof isn't leaking.

55. What is the union of set A and set B?

   Set A: {2,4,5,9,11}
   Set B: {3,5,8,11,13}

   [A] {2,3,4,5,5,8,9,11,11,13}

   [B] {2,3,4,5,8,9,11,13}

   [C] {5,11}

   [D] {2,3,4,8,9,13}

   [E] {5,9,13,20,24}

56. What is the intersection of set A and set B?

$$\text{Set A: } \{1,3,7,9,10,12,14\}$$
$$\text{Set B: } \{1,4,7,8,11,12,15\}$$

[A] {1,1,3,4,7,7,8,9,10,11,12,12,14,15}

[B] {1,3,4,7,8,9,10,11,12,14,15}

[C] {1,7,12}

[D] {1,1,7,7,12,12}

[E] {3,4,8,9,10,11,14,15}

57. Which statement is NOT implied by the Venn diagram below?

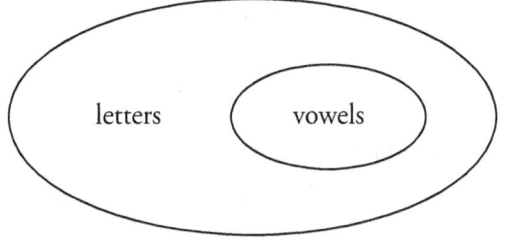

[A] No vowels are not letters.

[B] All vowels are letters.

[C] Some letters are vowels.

[D] Some letters are not vowels.

[E] Some vowels are not letters.

58. A total of 150 students have signed up for musical activities. There are 82 students in the choir and 80 students in the band. How many students are in both the band and the choir?

   [A] 12

   [B] 24

   [C] 42

   [D] 70

   [E] 162

59. Chris's older brother Mike is 2 years younger than Florence. When Tom's younger sister Rhoda was 8, Chris was 3. Florence is not older than Rhoda. Name the five people in ascending order of age.

   [A] Tom, Rhoda, Florence, Mike, Chris

   [B] Tom, Florence, Rhoda, Mike, Chris

   [C] Chris, Mike, Florence, Rhoda, Tom

   [D] Chris, Mike, Rhoda, Florence, Tom

   [E] Chris, Rhoda, Mike, Florence, Tom

60. Disprove the following statement by offering a counterexample:

   "Multiplying two numbers together produces a larger number than either of the two original numbers."

   [A] $\sqrt{2} \times \sqrt{2}$

   [B] 1.25×1.78

   [C] -3×-3

   [D] 0.5×0.6

   [E] -0.8×-0.3

Answer Key 1

Question Number	Correct Answer	Your Answer	Question Number	Correct Answer	Your Answer
1	B		31	E	
2	E		32	B	
3	B		33	C	
4	D		34	D	
5	E		35	A	
6	A		36	C	
7	B		37	C	
8	C		38	D	
9	C		39	E	
10	D		40	A	
11	B		41	B	
12	E		42	E	
13	B		43	C	
14	C		44	A	
15	C		45	D	
16	A		46	B	
17	C		47	C	
18	B		48	B	
19	A		49	E	
20	A		50	B	
21	E		51	E	
22	B		52	D	
23	C		53	D	
24	C		54	A	
25	E		55	B	
26	E		56	C	
27	D		57	E	
28	A		58	A	
29	A		59	C	
30	D		60	D	

Rationales for Test 1

The solutions presented represent one way to find the answer to the question.

1. Which of the following is closed under division?

    I. $\left\{\dfrac{1}{3}, 1, 3\right\}$

    II. $\{-1, 1\}$

    III. $\{-1, 0, 1\}$

    [A] I only

    [B] II only

    [C] III only

    [D] I and II

    [E] II and III

    The answer is B
    Set I is not closed under division, because $\dfrac{1}{3}$ divided by 3 is $\dfrac{1}{9}$, a number outside the set.

    Set III is not closed under division, because it is not possible to divide either −1 or 1 by 0.

2. Which of the following is always composite if x is an odd positive integer and y is an even positive integer greater than 1?

    [A] $x + y$

    [B] $|x + y|$

    [C] $x + 2y$

    [D] $3x + y$

    [E] $3xy$

The answer is E

$3xy$ must be composite, since 3, x, and y are all factors.

3. Find the LCM of 25, 18, and 24.

    [A] 1,200

    [B] 1,800

    [C] 2,400

    [D] 3,600

    [E] 10,800

The answer is B

The LCM must contain 2 factors of 5 to be a multiple of 25. It must contain 2 factors of 3 and a factor of 2 to be a multiple of 18. And it must contain 3 factors of 2 and a factor of 3 to be a multiple of 24. Therefore, the LCM must contain the following factors: $5 \times 5 \times 3 \times 3 \times 2 \times 2 \times 2 = 1800$

4. Solve for x: $|3x| + 6 = 21$

    [A] [9, −5]

    [B] [−9, 5]

    [C] [−5, 0, 5]

    [D] [−5, 5]

    [E] [−9, 9]

The answer is D

Write two equations to express the two possibilities:

$$3x + 6 = 21$$
$$-3x + 6 = 21$$

Solving the two equations gives 5 and −5.

5. Which graph represents the solution set for $x^2 - 5x > -6$?

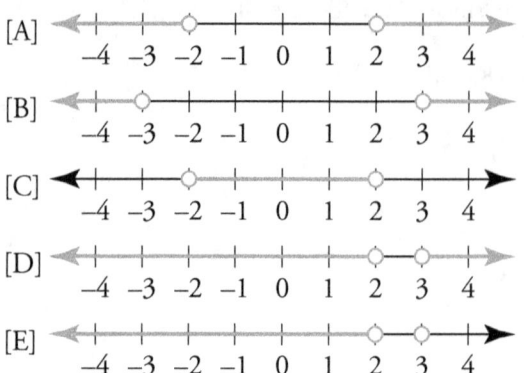

The answer is E

Gathering all terms on the left gives $x^2 - 5x + 6 > 0$. Replace the inequality symbol with an equals sign and solve for x: $x = 2$, and $x = 3$. A graph of the parabola makes clear that it is greater than 0 for x-values less than 2 or greater than 3 or greater, but less than 0 when $2 \leq x \leq 3$.

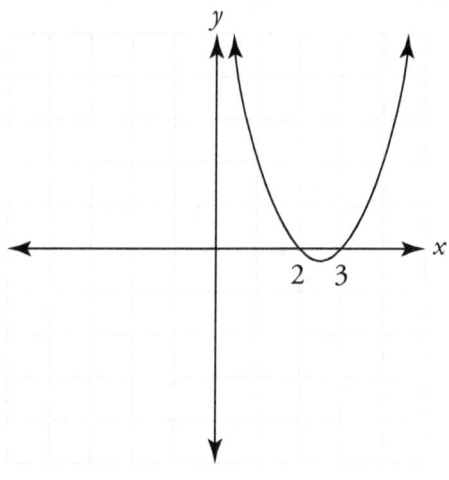

6. What is the equation of the graph shown below?

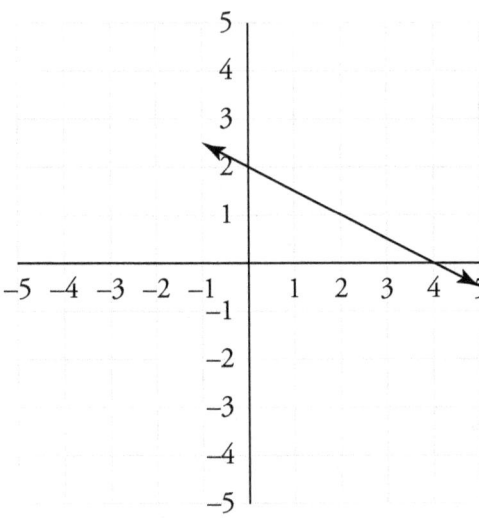

[A] $x + 2y = 4$

[B] $x - 2y = 4$

[C] $2x + y = 4$

[D] $x + 2y = -4$

[E] $x - 2y = -4$

The answer is A

Replacing x with 0 gives a y-intercept of 2. Replacing y with 0 gives an x-intercept of 4. The equation is linear, so a line can be drawn between the two points to complete the graph.

7. Solve the following inequality: $-2x > 4$

   [A] $x > -2$

   [B] $x < -2$

   [C] $x > 2$

   [D] $x > -8$

   [E] $x < 2$

The answer is B

To solve for x, you must divide by -2, but dividing by a negative number reverses the inequality, so the result is $x < -2$.

8. Which equation represents a circle centered on the origin with radius 3?

   [A] $x^2 + y^2 = 3$

   [B] $x^2 + y^2 = 6$

   [C] $x^2 + y^2 = 9$

   [D] $x^2 + y^2 = 36$

   [E] $x^2 - y^2 = 9$

The answer is C

The equation for a circle centered on the origin is $x^2 + y^2 = r^2$. Since $r = 3$, the equation in this case is $x^2 + y^2 = 9$.

9. Given that D is a distance, M is a mass, T is a time, and V is a velocity, which of the following units could be used to measure $\frac{MTV}{D}$?

   [A] feet

   [B] meters

   [C] grams

   [D] seconds

   [E] miles per hour

The answer is C

Try some sample units and see how they interact:

Let the distance be in miles, the mass be in grams, the time be in hours, and the velocity in miles per hour. Then the units to express $\frac{MTV}{D}$ would be $g \times h \times \frac{mi}{h} \times \frac{1}{mi}$. Hours and miles cancel out, leaving only grams.

10. Cubic meters are used to measure which of the following?

    [A] Distance

    [B] Length

    [C] Area

    [D] Volume

    [E] Mass

The answer is D

Distance and length are measured in linear meters. Area is measured in square meters. Mass is not measured in meters of any kind. Of the choices, only volume is measured in cubic meters.

11. What figure best describes a data set in which many items are clustered near the median value with a smaller number of values less than or greater than the median at greater distances on each side?

   [A] A parabola

   [B] A normal curve

   [C] A line of best fit

   [D] A Cartesian curve

   [E] A Newtonian curve

The answer is B

The figure described is a normal curve, called normal because data from the natural world tend to present a shape in which median values are commoner than extreme ones.

12. If you prove a theorem by showing that an attempt to prove the opposite of the theorem leads to a contradiction, you are using the logical strategy called:

   [A] Inductive reasoning

   [B] Exhaustive proof

   [C] Proof by attraction

   [D] Direct proof

   [E] Indirect proof

The answer is E

Such a proof is called "indirect" because it uses the opposite of the theorem instead of the theorem itself.

13. Compute the area of the shaded region, given a radius of 7 meters. Point O is the center.

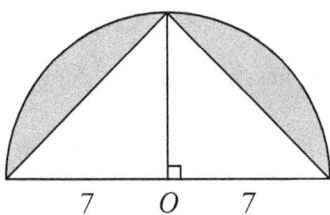

[A] 14.0

[B] 28.0

[C] 55.9

[D] 104.9

[E] 153.9

The answer is B

The area of the half circle is $\frac{49\pi}{2}$. The two triangles are equivalent to a square 7 meters on a side. So the shaded area $= \frac{49\pi}{2} - 49 \approx 28.0$.

14. A garden measures 25m by 40m including a circular fishpond with radius 3m. What is the area of the garden not including the fishpond?

[A] 101.7 m²

[B] 111.2 m²

[C] 971.7 m²

[D] 981.2 m²

[E] 990.6 m²

The answer is C

The area of the garden is $25 \times 40 = 1000$ m². The area of the fishpond is $3^2\pi \approx 28.3$ m². The difference is about 971.7 m².

15. The base of cone A has 3 times as great an area as the base of cone B but the height of cone A is only $\frac{1}{3}$ the height of cone B. Which statement is true?

[A] Cone A has 9 times the volume of cone B.

[B] Cone A has 3 times the volume of cone B

[C] Cone A and cone B have the same volume.

[D] Cone B has 3 times the volume of cone A.

[E] Cone B has 9 times the volume of cone A.

The answer is C

Let h be the height of cone B and let b be the area of the base of cone B. Using the formula for the volume of a cone, the volume of cone B is $\frac{1}{3}bh$. The base of cone $A = 3b$, while the height of cone $A = \frac{h}{3}$. Therefore, the volume of cone A is $\frac{1}{3}(3b)\left(\frac{h}{3}\right) = \frac{1}{3}bh$, the same as cone B.

16. Find the area of the figure depicted below.

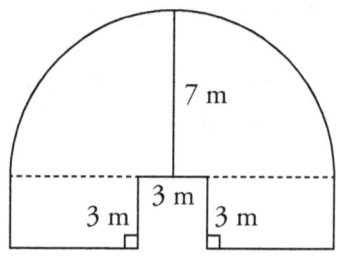

[A] 109.9 m²

[B] 118.9 m²

[C] 142.9 m²

[D] 144.9 m²

[E] 186.9 m²

The answer is A

The area of the circle is $\frac{49\pi}{2}$. The dotted line equals a diameter, twice the length of the radius, or 14. Subtracting the gap of 3 m, the two rectangles add up to a length of 11 m times a width of 3 m So the total area is $\frac{49\pi}{2} + 33 \approx 109.9$ m^2.

17. State the domain of the function $f(x) = \frac{2x-14}{x^2-9}$.

   [A] $x \neq 3$

   [B] $x \neq 3, 7$

   [C] $x \neq 3, -3$

   [D] $x \neq 7$

   [E] $x = 3, -3, 7$

The answer is C

The domain must exclude values of x that would cause the denominator of the function to equal 0. Therefore, both −3 and 3 are excluded from the domain.

18. Which of the following is a factor of the expression $6x^2 - 5x - 14$?

   [A] $3x + 7$

   [B] $6x + 7$

   [C] $6x - 7$

   [D] $6x - 5$

   [E] $x + 2$

The answer is B

To factor the expression, multiply 6 times 14 to get 84. Then look for two factors of 84 that differ by 5: 7 and 12. Use these factors to rewrite the middle term as $7x - 12x$. You can then factor the expression as $(6x + 7)(x - 2)$.

19. Solve for x by factoring: $x^2 + x - 6 = 0$

   [A] $x = (-3, 2)$

   [B] $x = (3, -2)$

   [C] $x = (-6, 1)$

   [D] $x = (6, -1)$

   [E] no real solutions

The answer is A

Factoring the left side of the equation gives us $(x + 3)(x - 2) = 0$. Setting each factor equal to 0 gives us solutions of -3 and 2.

20. Which of the following is equivalent to $\sqrt[b]{x^a}$?

   [A] $x^{\frac{a}{b}}$

   [B] $x^{\frac{b}{a}}$

   [C] $a^{\frac{x}{b}}$

   [D] $b^{\frac{x}{a}}$

   [E] $a^{\frac{b}{x}}$

The answer is A

Taking the bth root of x^a is equivalent to dividing the exponent of x^a by b.

21. Given $f(x) = 2x + 1$ and $g(x) = x^2 - 1$, determine $g(f(x))$.

[A] $4x^2 + 4x - 1$

[B] $4x^2 + 4x + 1$

[C] $4x^2$

[D] $4x^2 - 1$

[E] $4x^2 + 4x$

The answer is E
If $f(x) = 2x + 1$, $g(f(x)) = (2x + 1)^2 - 1 = 4x^2 - 4x$.

22. Compute the median for the following data set: {9, 11, 18, 13, 12, 21}

[A] 12

[B] 12.5

[C] 13

[D] 14

[E] 15.5

The answer is B
In ascending order, the set is {9, 11, 12, 13, 18, 21}

Since there are an even number of data items, the median is halfway between the two most central items when the items are put in ascending order, in this case the third and fourth.

23. Which graph represents the equation $y = x^2 + 3x$?

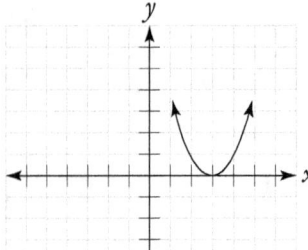

[A]

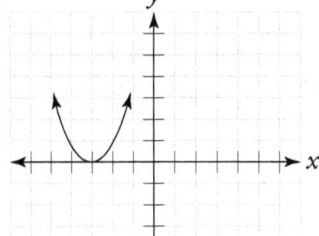

[B]

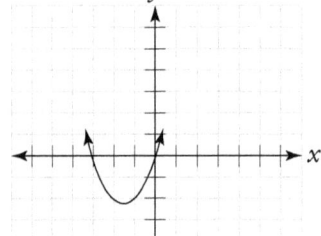

[C]

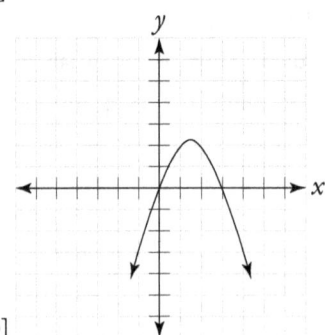

[D]

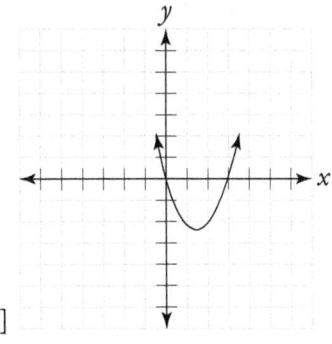

[E]

The answer is C

Since $x^2 + 3x$ can be factored as $x(x + 3)$, the function has zeroes at 0 and −3. Since the first term is positive, the parabola opens up. Choice C fits these specifications.

24. What would be the best measure of central tendency for the following collection of high temperatures on 10 successive days? {27, 24, 33, 24, 36, 65, 34, 30, 28, 29}

   [A] Mean

   [B] Either mean or median

   [C] Median

   [D] Mode

   [E] Either median or mode

The answer is C

Since the data contains an outlier, the mean would be skewed too high. The mode is the smallest data item and therefore also not a good representation. The median is the best available representation of the data as a whole.

25. If the correlation between two variables is zero, the association between the two variables is

   [A] Negative linear

   [B] Positive linear

   [C] Quadratic

   [D] Direct variation

   [E] Random

The answer is E

Choices A, B, C, and D all describe some form of correlation between the two variables. Only a random association shows zero correlation.

26. Which of the following is not a valid method of collecting statistical data?

   [A] Random sampling

   [B] Systematic sampling

   [C] Volunteer response

   [D] Weighted sampling

   [E] Cylindrical sampling

The answer is E

Choices A, B, C, D describe methods of data collection with varying degrees of potential usefulness and prohibition. There is no such thing as cylindrical sampling.

27. A jar contains 3 red marbles and 7 green ones. What is the probability that a marble picked at random from the jar will be red?

[A] $\dfrac{1}{3}$

[B] $\dfrac{1}{7}$

[C] $\dfrac{3}{7}$

[D] $\dfrac{3}{10}$

[E] $\dfrac{7}{10}$

The answer is D

Three marbles are red out of a total of 10 marbles, yielding a probability of 3/10.

28. A die is rolled several times. What is the probability that a 6 will not appear before the fourth roll of the die?

[A] $\dfrac{125}{216}$

[B] $\dfrac{625}{1296}$

[C] $\dfrac{1}{2}$

[D] $\dfrac{5}{6}$

[E] $\dfrac{1}{216}$

The answer is A

Each time the die is rolled, the chance of rolling a number other than 6 is $\dfrac{5}{6}$. The probability that this will happen three times is $\dfrac{5}{6} \times \dfrac{5}{6} \times \dfrac{5}{6} = \dfrac{125}{216}$.

29. There is a 30% chance of rain this Saturday and a 30% chance of rain on Sunday as well. What is the chance of rain on both days?

[A] 9%

[B] 30%

[C] 49%

[D] 60%

[E] 70%

The answer is A

The probability of two things both happening is the product of the two probabilities: $0.3(0.3) = 0.09 = 9\%$.

30. Which equation matches the data in the table?

x	3	4	5	6
y	7	8	9	10

[A] $y = 2x - 1$

[B] $y = 2x + 1$

[C] $y = -x + 10$

[D] $y = x + 4$

[E] $y = x - 4$

The answer is D

Each y-value is 4 greater than the corresponding x-value.

31. Which table could be generated by the equation: $y = x^2 + 2x - 1$?

[A]
x	1	2	3	4
y	2	5	8	11

[B]
x	1	2	3	4
y	4	9	16	25

[C]
x	1	2	3	4
y	1	5	11	19

[D]
x	1	2	3	4
y	2	7	13	21

[E]
x	1	2	3	4
y	2	7	14	23

The answer is E

Substitute *each* x-value into the equation and see if the result matches the y-value. Only in table E do all the y-values correspond to the values found by substituting the x-values into the equation.

32. The fees charged by a parking garage are as follows:

Hours	1	2	3	4	55
Fee	$12	$19	$26	$33	$40

How would you summarize the fees charged?

[A] $12 an hour

[B] $5 plus $7 per hour

[C] $15 an hour with a $3 discount

[D] $4 plus $8 per hour

[E] $3 plus $9 per hour

The answer is B

Each additional hour costs $7 more, so the rate must be $7 an hour, which leaves $5 as the initial fee.

33. Which of the following is a solution to $x^2 + 4x + 4 = 25$?

    [A] 2

    [B] –2

    [C] –7

    [D] –3

    [E] 5

The answer is C

Taking the square root of both sides yields $x + 2 = \pm 5$. Therefore, $x = 3$ or –7.

34. Solve the following system of equations:
$$2x + y = 8$$
$$4x + 2y = 20$$

    [A] $x = 2, y = 4$

    [B] $x = 3, y = 1$

    [C] $x = 4, y = 0$

    [D] no solutions

    [E] an infinite number of solutions

The answer is D

Multiply the first equation by 2 and subtract from the second equation. The result is 0 = 4. A system of equations that resolves to an untrue statement has no solutions.

35. If an initial deposit of $10,000 is made to a savings account with interest compounded continuously at an annual rate of 6% how much money is in the account after 5 years?

    [A] $13,498.59

    [B] $3498.59

    [C] $13,382.26

    [D] $3,382.26

    [E] $13,000.00

The answer is A

Continuously compounded interest is calculated using the formula Pe^{rt}, where P is the amount of the principal, r is the annual rate, and t is the time in years.
$$10,000 \times e^{0.06 \times 5} = 10,000 e^{0.3} \approx 13,498.59.$$

36. A dance team comes prepared with a tango, a waltz, a disco number, a salsa routine, and a ballet selection. In how many different orders can they present their routines?

    [A] 5

    [B] 25

    [C] 120

    [D] 625

    [E] 3125

The answer is C

Any of the 5 routines could be the first number. The second number could be any of the remaining 4, the third could be any of the remaining 3, and so on. The total number of choices is $5 \times 4 \times 3 \times 2 \times 1 = 120$.

37. You can choose 3 selections from a buffet table with 8 dishes. How many different plates can you choose?

    [A] 6

    [B] 24

    [C] 56

    [D] 336

    [E] 6561

The answer is C

Since the order of items on your plate does not matter, it is combinations rather than permutations we need to find. The number of combinations of k items out of n possible selections is given by the formula $\frac{n!}{(n-k)!k!} \cdot \frac{8!}{5!3!} = 56$.

38. Leah has 4 blouses, 3 skirts, and 6 pairs of shoes. How many different outfits can she dress herself in?

    [A] 12

    [B] 13

    [C] 24

    [D] 72

    [E] 720

The answer is D

By the Fundamental Counting Principle, the number of different outfits is $4 \times 3 \times 6 = 72$.

39. Hiroshi surveys his classmates to find what percent of them come to school on the bus, by car, by subway, by bicycle, or on foot. What is the best way to display his results?

[A] A line graph

[B] A box plot

[C] A stem-and-leaf plot

[D] A scatterplot

[E] A circle graph

The answer is E

A circle graph is the best way to display what portion of the whole data set is occupied by each item.

40. Which equation could be used as a line of best fit for the scatterplot below?

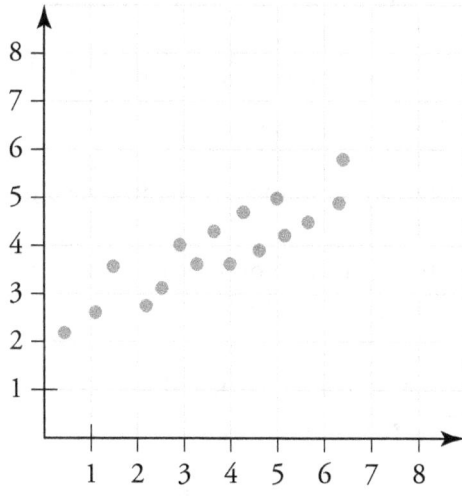

[A] $y = \dfrac{1}{2}x + 2$

[B] $y = 2x + 2$

[C] $y = -2x + 2$

[D] $y = \dfrac{1}{2}x - 2$

[E] $y = \dfrac{1}{2}x + 2$

The answer is A

The data appear to start around 2, and the *y*-values are generally rising less fast than the *x*-values, so the slope appears to be less than 1. $\dfrac{1}{2}x + 2$ is the best fit among the choices.

41. To find the standard variation of a data set, you first compute the square of the distance of each data item from the mean of all the data items. Then what do you do?

    [A] Add all the squared distances and take the square root of the result.

    [B] Find the mean of the squared distances and take the square root of the result.

    [C] Multiply the squared distances and take the nth root of the result.

    [D] Multiply the square root of the sum of the squared distances by the mean of the squared distances.

    [E] Multiply the sum of the squared distances by the square root of the mean of the squared distances.

The answer is B

Choice B correctly completes the process of finding a standard variation.

42. In which data set is the mode greater than the median?

   [A] {9,11,11,12,14}

   [B] {13,15,17,19,21]

   [C] {8,11,12,12,19}

   [D] {9,9,9,14,20}

   [E] {7,11,13,14,14}

The answer is E

In choice E, the median is 13 and the mode 14.

43. Of the 200 students in the junior class, 8% are in the Spanish Club. How many juniors are in the Spanish Club?

   [A] 4

   [B] 8

   [C] 16

   [D] 20

   [E] 25

The answer is C

The number of juniors in the Spanish club is 8 % of 200 or 16.

44. When Olga bought a boat for $1,750 she paid an excise tax of $78.75. What was the percent of the tax?

[A] 4.5%

[B] 5.5%

[C] 6.3%

[D] 7%

[E] 7.5%

The answer is A

To find the percent of tax, divide the tax by the sales price and multiply by 100. $\frac{78.75}{1750} \times 100 = 4.5$, so the tax rate is 4.5%.

45. A bank account pays 5% interest yearly. How large an amount would have to be deposited to earn $75 interest in a year?

[A] $ 375

[B] $875

[C] $1200

[D] $1500

[E] $3750

The answer is D

If a principal of x dollars earns $75 at 5% interest, then $0.5x = 75$. Multiplying both sides by 20 yields $x = 1500$, so the amount of the principal must be $1500.

46. A stock previously trading at $96 a share is now trading at $88 a share. What is the percent of change in the value of the stock?

[A] −8%

[B] −8.3%

[C] −12%

[D] −12.5%

[E] −16%

The answer is B

The percent of change is found by dividing the amount of the change by the original value, then multiplying by 100. The change is −$8, and the original amount is $96. $\frac{-8}{96} \times 100 \approx -8.3$, so the stock price has changed about −8.3%.

47. The admission price to tour the Haunted House has been changed from $25 to $30 What is the percent of change in the admission price?

[A] 5%

[B] 16.7%

[C] 20%

[D] 25%

[E] 30%

The answer is C

The amount of change is +$5, and the original value is $25. $\frac{5}{25} = \frac{1}{5} = 20\%$.

48. Eileen's Bakery had expenses of $62,500 last year and sales of $68,750. What was the profit as a percent of the expenses?

   [A] 6.25%

   [B] 10%

   [C] 12%

   [D] 15%

   [E] 16.7%

The answer is B
   The amount of change is $6,250. $\frac{6250}{62500} = \frac{1}{10} = 10\%$.

49. Tim's Typewriters had expenses of $26,200 last year and sales of $19,912. What was the loss as a percent of the expenses?

   [A] 7%

   [B] 8%

   [C] 16.7%

   [D] 20%

   [E] 24%

The answer is E
   The amount of loss was $6288, and 6288./26200 = 24%.

50. A stock that had been selling at $30 a share increased its share price by 20%. Later in the day the same stock suffered a 20% decrease in its share price. What was the price at the end of the day?

   [A] $24

   [B] $28.80

   [C] $30

   [D] $33

   [E] $36

The answer is B

After the $30 price increased by 20%, it was $36.

51. A sweater is marked "25% off." The sale price is $36. What was the price before the discount?

   [A] $27

   [B] $32

   [C] $40

   [D] $45

   [E] $48

The answer is E

If the original price has been decreased by 25%, the sale price is 75% of the original. Solving $36 = 0.75x$ yields $x = 48$.

52. The sum of $1440 is deposited in a bank which pays 6% simple interest per year. After how many years will there be $1872 in the account?

   [A] 2.5 years

   [B] 3 years

   [C] 4 years

   [D] 5 years

   [E] 8 years

The answer is D

After each year is completed, the amount in the account is increased by 0.06(1440) = $86.40 dollars. The number of years required to bring the account to $1872 is $\frac{1872-1440}{86.40} = \frac{432}{86.40} = 5$

53. A bank pays 5% interest on deposits, compounded yearly. If $14,000 is deposited, how much will be in the account 3 years later?

   [A] $14,350

   [B] $15,435

   [C] $16,100

   [D] $16,206.75

   [E] $17,500

The answer is D

The amount in the account after 3 years will be $14,000 \times 1.05^3$ = $16,206.75.

54. Which statement is logically equivalent to the following: If it's raining, my roof is leaking.

[A] If my roof isn't leaking, it isn't raining.

[B] If my roof is leaking, it's raining.

[C] If it isn't raining, my roof isn't leaking.

[D] If my roof is leaking, it's not raining

[E] If it's raining, my roof isn't leaking.

The answer is A

The contrapositive of a true statement is also true. In this case, rain always makes my roof leak, so the absence of a leak could only be explained by the absence of rain.

55. What is the union of set A and set B?

$$\text{Set A: } \{2,4,5,9,11\}$$
$$\text{Set B: } \{3,5,8,11,13\}$$

[A] {2,3,4,5,5,8,9,11,11,13}

[B] {2,3,4,5,8,9,11,13}

[C] {5,11}

[D] {2,3,4,8,9,13}

[E] {5,9,13,20,24}

The answer is B

The union of the two sets contains every number that is in either set. Numbers that are in both sets are included only once in the union set.

56. What is the intersection of set A and set B?

   Set A: {1,3,7,9,10,12,14}
   Set B: {1,4,7,8,11,12,15}

   [A] {1,1,3,4,7,7,8,9,10,11,12,12,14,15}

   [B] {1,3,4,7,8,9,10,11,12,14,15}

   [C] {1,7,12}

   [D] {1,1,7,7,12,12}

   [E] {3,4,8,9,10,11,14,15}

The answer is C

The intersection of the two sets contains only those numbers that are in both sets.

57. Which statement is NOT implied by the Venn diagram below?

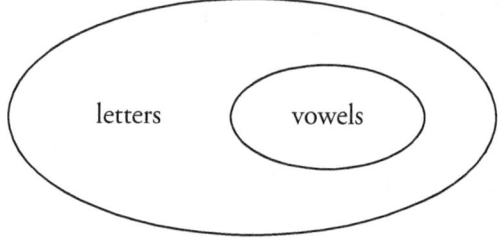

   [A] No vowels are not letters.

   [B] All vowels are letters.

   [C] Some letters are vowels.

   [D] Some letters are not vowels.

   [E] Some vowels are not letters.

The answer is E

The diagram shows that the class of vowels is totally included in the class of letters, so there are no vowels that are not letters.

58. A total of 150 students have signed up for musical activities. There are 82 students in the choir and 80 students in the band. How many students are in both the band and the choir?

[A] 12

[B] 24

[C] 42

[D] 70

[E] 162

The answer is A

The sum of 82 choir members and 80 band members is 162, but only 150 students are involved. The explanation is that 12 students are in both the band and choir, and are therefore counted twice when the two memberships are added.

59. Chris's older brother Mike is 2 years younger than Florence. When Tom's younger sister Rhoda was 8, Chris was 3. Florence is younger than Rhoda. Name the five people in ascending order of age.

[A] Tom, Rhoda, Florence, Mike, Chris

[B] Tom, Florence, Rhoda, Mike, Chris

[C] Chris, Mike, Florence, Rhoda, Tom

[D] Chris, Mike, Rhoda, Florence, Tom

[E] Chris, Rhoda, Mike, Florence, Tom

The answer is C

Statement 1 allows as to put Chris, Mike, and Florence in ascending order. The second statement allows us to put Rhoda and Tom in ascending order. Since Florence is younger than Rhoda, both Rhoda and Tom are older than Chris, Mike, and Florence, allowing us to put all five in ascending order.

60. Disprove the following statement by offering a counterexample:

"Multiplying two numbers together produces a larger number than either of the two original numbers."

[A] $\sqrt{2} \times \sqrt{2}$

[B] 1.25×1.78

[C] -3×-3

[D] 0.5×0.6

[E] -0.8×-0.3

The answer is D

The product of 0.5 and 0.6 is 0.3, which is smaller than either of the two original numbers. The other multiplications produce a product larger than either of the numbers multiplied.

Sample Test 2

Sample Test Questions

Directions: Read each item and select the best response.

1. Solve for x: $\sqrt{3x} + 4 = x - 2$

   [A] 3

   [B] 9

   [C] 12

   [D] Both A and C

   [E] None of the above

2. Solve for x: $7 - 5x = 7x - 11 - 3x$

   [A] −2

   [B] 2

   [C] 0

   [D] $-1\frac{1}{2}$

   [E] $3\frac{1}{4}$

3. Find the zeroes of $f(x) = x^3 + x^2 - 14x - 24$

   [A] 4, 3

   [B] 3, −8

   [C] 7, −2, −1

   [D] 4, 3, 2

   [E] 4, −3, −2

4. Expand $(x + 2)^3$

   [A] $6x$

   [B] $3x + 6$

   [C] $x^3 + 8$

   [D] $x^3 + 3x^2 + 8$

   [E] $x^3 + 6x^2 + 12x + 8$

5. Which of the following sets of ordered pairs does not represent a function?

   [A] {(1,4), (2,5), (3,6)}

   [B] {(1,–1), (2,–2), (3,–3)}

   [C] {(3,1), (4,1), (5,1)}

   [D] {(1,3), (1,4), (1,5)}

   [E] All of the above do represent functions

6. Solve for x $\dfrac{2}{x} + \dfrac{3}{8} = \dfrac{5}{2x}$

   [A] $\dfrac{4}{3}$

   [B] $\dfrac{8}{3}$

   [C] 4

   [D] 8

   [E] 16

7. Solve for y: $4x - 3y = 9$

   [A] $y = 4x - 9$

   [B] $y = 4x + 3$

   [C] $y = 3 - \frac{4}{3}x$

   [D] $y = \frac{4}{3}x - 3$

   [E] $y = \frac{5}{3}x$

8. State the domain of the function $f(x) = \frac{3x-6}{x^2-25}$

   [A] $x \neq 2$

   [B] $x \neq 5, -5$

   [C] $x \neq 2, -2$

   [D] $x \neq 5$

   [E] All real numbers

9. Solve $9x^2 - 25 = 0$

   [A] $\pm \frac{5}{3}$

   [B] ± 5

   [C] 0

   [D] $\pm \frac{5i}{3}$

   [E] ± 4

10. What is the equation of the graph shown below?

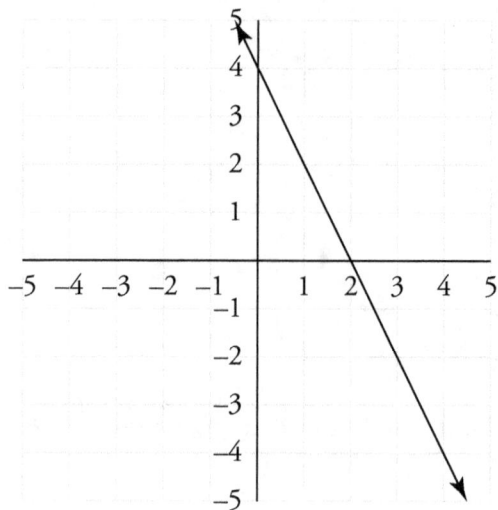

[A] $x + 2y = 4$

[B] $x - 2y = 4$

[C] $2x + y = 4$

[D] $x + 2y = -4$

[E] $x - 2y = -4$

11. Find the reflection of the point (4,7) over the y axis.

[A] (7,4)

[B] (4,–7)

[C] (0,7)

[D] (–4,7)

[E] (4, 0)

12. If $Pv = nrt$, solve for n.

    [A] $n = \dfrac{Pv}{rt}$

    [B] $n = Pv - rt$

    [C] $n = Pvrt$

    [D] $n = P + v - r - t$

    [E] none of the above

13. If $f(x) = 10 - 3x^2$, find $f(-5)$

    [A] −215

    [B] −65

    [C] 85

    [D] 175

    [E] 235

14. In a class of 24 students, 3 children are randomly chosen to form a focus group. How many different groups can be made?

    [A] 8

    [B] 72

    [C] 2,024

    [D] 12,144

    [E] 13,824

15. Given a spinner marked as shown, what is the probability of spinning a multiple of 6 or a multiple of 4?

[A] $\dfrac{7}{16}$

[B] $\dfrac{5}{16}$

[C] $\dfrac{1}{2}$

[D] $\dfrac{1}{8}$

[E] $\dfrac{3}{8}$

16. Given a standard deck of cards, a king is drawn and placed face up on the table. What is the probability that a second king will be randomly chosen in a second draw?

[A] $\dfrac{1}{13}$

[B] $\dfrac{4}{13}$

[C] $\dfrac{3}{52}$

[D] $\dfrac{3}{51}$

[E] $\dfrac{1}{3}$

17. If P(R) = 0.6, what is the P(not R)?

   [A] 1

   [B] 0.6

   [C] 0.4

   [D] 0.3

   [E] 0

18. If A and B are independent events, and P(A) = 20% and P(B) = 70%, find P(A ∩ B).

   [A] 90%

   [B] 50%

   [C] 35%

   [D] 14%

   [E] 5%

19. A spinner contains sections for the numbers 1–12 but the multiples of three appear twice. What is the expected value of a spin?

[A] 1

[B] 3

[C] 6

[D] 6.75

[E] 10.5

20. In the first four games of the season, a basketball team scores 25, 30, 33, and 28 points per game. How many points does the team need to score in their 5th game to reach an average of 30 points per game?

   [A] 29

   [B] 30

   [C] 32

   [D] 33

   [E] 34

21. Find the mode for the given set of data. {1, 12, 22, 32, 32}

   [A] 2

   [B] 22

   [C] 27

   [D] 31

   [E] 32

22. Find the range for the given set of data. {1, 12, 22, 32, 32}

    [A] 1

    [B] 22

    [C] 27

    [D] 31

    [E] 32

23. If the median test score in a class is 88%, which of the following statements correctly follow?

    [A] The teacher grades using the Normal Distribution.

    [B] The mean is also 88%.

    [C] The same number of students scored above 88% as scored below 88%.

    [D] The range of test scores is 60–100%.

    [E] None of the above.

24. Which of the following represents a true statement?

    [A] The range of a set of data represents the numerical distance between the smallest and largest pieces of data.

    [B] A standard deviation is the same as a quartile.

    [C] A set of data can have only one outlier

    [D] A set of data can have only one mode

    [E] The mean and median of a set of data are the same value.

25. Which of the following is not a conclusion that can be drawn from the bar graph below?

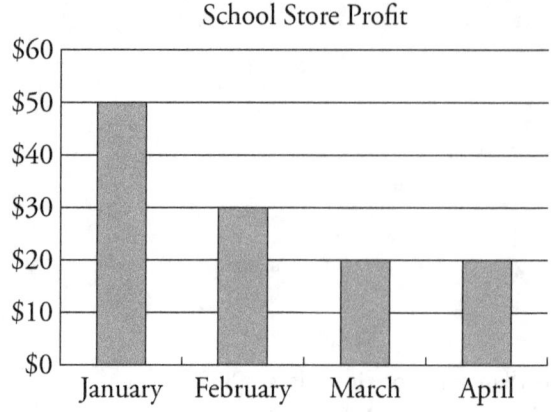

[A] Profit for December was approximately between $50 and $60.

[B] Profits for March and April were approximately equal.

[C] Profits decreased since January.

[D] The average profit is approximately $30 per month.

[E] Both A and D are invalid conclusions.

26. The phrase that best describes this graph of data is

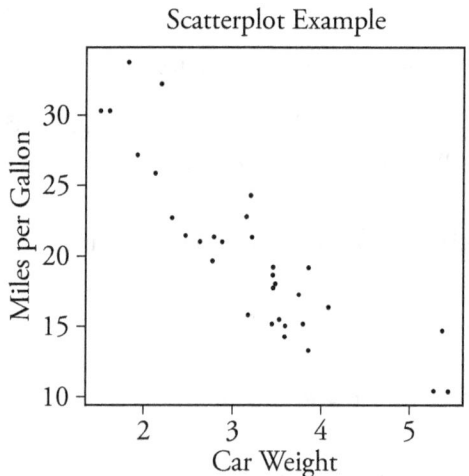

[A] Positively Correlated

[B] Negatively Correlated

[C] Exponentially Correlated

[D] Continuously Correlated

[E] Not Correlated

27. Which of the following statements is true regarding the graph of a set of data that is normally distributed?

    [A] The maximum value of the graph is the same as the mean, median, and mode of the set of data.

    [B] The standard deviation is ±3

    [C] The graph crosses the x axis at the 5th standard deviation from the mean.

    [D] All of the above

    [E] None of the above

28. 60 is what percent of 400?

    [A] 10%

    [B] 15%

    [C] 24%

    [D] 25%

    [E] 69%

29. A customer buys a sweater for $24, pants for $28, and socks for $4. Find the total amount the customer will be charged if the sales tax is 8.5%.

   [A] $4.76

   [B] $47.60

   [C] $56.55

   [D] $60.76

   [E] $64.50

30. If the price of a car rises from $15,900 to $17,000, find the percent increase.

   [A] 6.5%

   [B] 6.9%

   [C] 9.4%

   [D] 11%

   [E] 25%

31. A 15% service charge was added to labor costs making the final billed amount $275. How much was the cost of labor?

   [A] $220

   [B] $234

   [C] $239

   [D] $259

   [E] $260

32. If a credit card company applies an APR of 22% to its monthly statements, find the effective interest rate charged.

   [A] 10%

   [B] 22.2%

   [C] 23.8%

   [D] 24.4%

   [E] 25%

33. $2,000 is invested in an account paying 3.7% annual interest compounded quarterly. Find the balance of the account after 10 years.

   [A] $2,037.04

   [B] $2,740

   [C] $2,890.55

   [D] $2,900

   [E] $3,010.25

34. Find the balance, after 8 years, when $2,000 is deposited in an account with an APR of 5% compounded continuously.

   [A] $2,040

   [B] $2,954.91

   [C] $2,983.65

   [D] $2,800

   [E] $2,851.23

35. Find the present value needed to achieve a future value, in 15 years' time, of $10,000 if the investment earns an APR of 8% compounded annually.

   [A] $2,500

   [B] $2,815

   [C] $3,000

   [D] $3,129

   [E] $3,152

36. A customer is choosing between 3 different interest bearing accounts.
   - Choice 1 offers an APR of 6% compounded quarterly.
   - Choice 2 offers an APR of 5.5% compounded monthly.
   - Choice 3 offers an APR of 5% compounded weekly.

   Which will produce a higher future value over a set period of time?

   [A] Choice 1

   [B] Choice 2

   [C] Choice 3

   [D] They all offer the same future value

   [E] There is not enough information given to determine

37. Given $\triangle DOG$ is isosceles with $\overline{DO} \cong \overline{OG}$ and m$\angle$O = 22°, find m$\angle$D.

   [A] 180°

   [B] 158°

   [C] 79°

   [D] 22°

   [E] none of the above

38. Find the surface area of a box which is 3 feet wide, 5 feet tall, and 4 feet deep.

   [A] 47 sq. ft.

   [B] 60 sq. ft.

   [C] 94 sq. ft

   [D] 188 sq. ft.

   [E] 200 sq. ft.

39. Based on the diagram below

   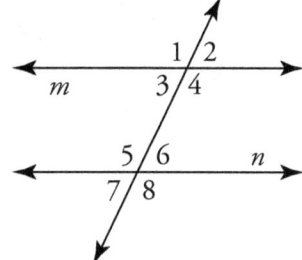

   If line *m* is parallel to line *n*, then which pair of angles is congruent?

   [A] 1 and 2

   [B] 3 and 5

   [C] 2 and 8

   [D] 2 and 7

   [E] No pair of angles is congruent

40. Which of the following is true about a parallelogram?

   [A] The opposite sides are congruent.

   [B] The diagonals are congruent.

   [C] All four angles are congruent.

   [D] The sum of the interior angles is 180°.

   [E] All of the above

41. Which theorem can be used to prove $\triangle BAT = \triangle MKT$?

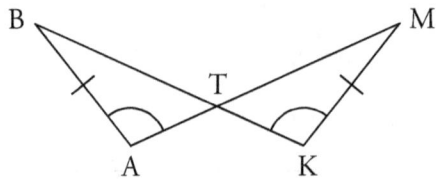

   [A] SSS

   [B] ASA

   [C] SAS

   [D] AAS

   [E] CPCTC

42. Which of the following is *not* true? An equilateral triangle:

   [A] has three equal sides

   [B] has three equal angles

   [C] has three acute angles

   [D] has three obtuse angles

   [E] has 3 vertices

43. An architect who wants to increase the area of a square room by a factor of 100 should

   [A] Make each side of the room 100 times longer

   [B] Make each side of the room 10 times longer

   [C] Double the height of the ceiling

   [D] Make the room rectangular

   [E] Make no change in the dimensions

44. If quadrilateral *ABCD* is congruent to *PQRS*, then which of the following is true?

   [A] $\overline{AD} \cong \overline{PS}$

   [B] $\overline{AB} \cong \overline{QR}$

   [C] $\angle A \cong \angle S$

   [D] $\angle B \cong \angle P$

   [B] None of the above

45. Given A = {H, O, R, S, E} and B = {C, O, W}, find A ∩ B.

   [A] {O}

   [B] { H, O, R, S, E, C, O, W}

   [C] {P, I, G}

   [D] {H, C}

   [E] ∅

46. Which statement below is true, based on the given Venn diagram?

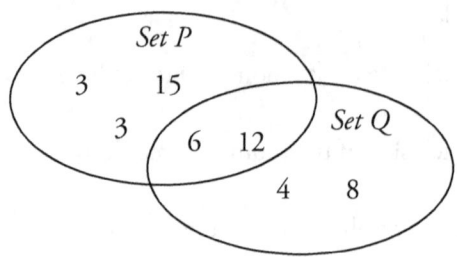

[A] P ⊂ Q

[B] P = 2Q

[C] 8 ∈ P

[D] P ∩ Q = {6, 12}

[E] None of the above

47. Given the conditional statement: "If you heat water to 212°F, it boils," find the converse.

[A] If you chill water to 32°F, it freezes.

[B] If water is boils, it is 212°F.

[C] If you don't heat water to 212°F, then it won't boil.

[D] If you boil water, it turns to steam.

[E] None of the above.

48. Which statement below represents a valid conclusion?

   [A] If n, then p. If p, then w. Therefore if n, then w.

   [B] If b, then q. Therefore if q, then b.

   [C] If k, then t or x. Not x. Therefore, if k, then t.

   [D] All of the above are valid.

   [E] Both A and C are valid.

49. Choose the most appropriate conclusion, based on the following statements;
   - Every rhombus is a parallelogram
   - Every square is a rhombus and a rectangle.
   - A trapezoid is not a parallelogram
   - Every parallelogram is a quadrilateral.

   Conclusion:

   [A] A parallelogram is a square.

   [B] A trapezoid is not a rhombus.

   [C] A rectangle is a square.

   [D] A parallelogram is a rhombus.

   [E] A quadrilateral is a rhombus.

50. Which statement below is true, based on the given Venn diagram?

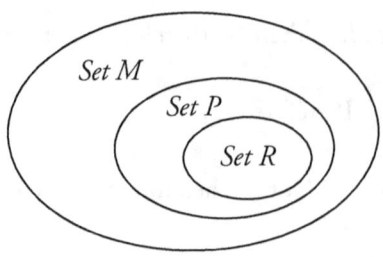

[A] R ⊂ M

[B] M ⊂ R

[C] P ∪ R = M

[D] R ∩ P = ∅

[E] None of the above

51. The Venn diagram below represents sports played by boys in a 6th grade class. Which statement below is true, based on the diagram?

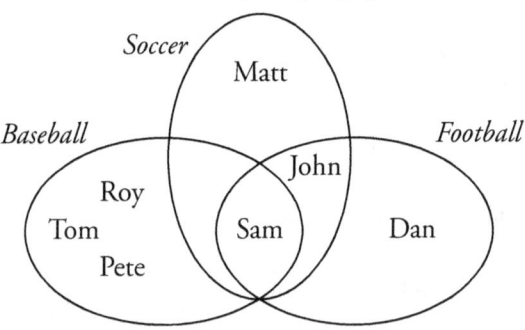

[A] Matt plays all 3 sports.

[B] 3 boys play baseball.

[C] John plays both football and soccer.

[D] Andy plays football.

[E] None of the above.

52. Given M = {1, 2, 3} and P = {3, 4}, find M × P.

   [A] {1, 2, 3, 4}

   [B] {1,4}

   [C] {(1, 3), (2, 4)}

   [D] {(1, 3), (1, 4), (2, 3), (2, 4), (3, 3), (3, 4)}

   [E] ∅

53. Which notation below represents the data graphed on the following number line?

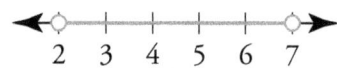

   [A] {2, 7}

   [B] {3, 4, 5, 6}

   [C] $\{x \mid 2 < x < 7\}$

   [D] $\{x \mid x > 2 \text{ and } x < 7\}$

   [E] Both C and D

54. Simplify $\dfrac{|8-40|+12}{2}$

   [A] 20

   [B] 22

   [C] 24

   [D] 30

   [E] 38

55. Which expression below is divisible by 6?

   [A] $x + 6$

   [B] $6x$

   [C] $3x(2y)^4$

   [D] All of the above

   [E] B and C

56. The Commutative Property applies to which of the following operations?

   [A] Addition

   [B] Multiplication

   [C] Division

   [D] All of the above

   [E] A and B

57. Amy needs 18 ft of rope. The rope costs $2.75 per yard. How much should she expect to pay?

   [A] $2.75

   [B] $16.50

   [C] $20.75

   [D] $49.50

   [E] $57.75

58. Which of the following is an irrational number?

   [A] $\dfrac{1}{3}$

   [B] −0.6

   [C] $\sqrt{36}$

   [D] $\sqrt{37}$

   [E] All of the above are irrational numbers

59. Calculate $(2.4 \times 10^8)(4.7 \times 10^{10})$

   [A] 11.28×10^{80}

   [B] 1.128×10^{81}

   [C] 1.128×10^{19}

   [D] 1.128×10^{18}

   [E] 11,280,000,000

60. If 2 cups of chocolate chips are used in a recipe to make 60 cookies, how many cups of chocolate chips will be needed to make 210 cookies?

   [A] 4 cups

   [B] 7 cups

   [C] 8 cups

   [D] 10 cups

   [E] 30 cups

Answer Key 2

Question Number	Correct Answer	Your Answer	Question Number	Correct Answer	Your Answer
1	C		31	C	
2	B		32	D	
3	E		33	C	
4	E		34	C	
5	D		35	E	
6	A		36	A	
7	D		37	C	
8	B		38	C	
9	A		39	D	
10	C		40	A	
11	D		41	D	
12	A		42	D	
13	B		43	B	
14	C		44	A	
15	E		45	A	
16	D		46	D	
17	C		47	B	
18	D		48	E	
19	D		49	B	
20	E		50	A	
21	E		51	C	
22	D		52	D	
23	C		53	E	
24	A		54	B	
25	A		55	E	
26	B		56	E	
27	A		57	B	
28	B		58	D	
29	D		59	C	
30	B		60	B	

Rationales for Test 2

The solutions presented represent one way to find the answer to the question.

1. Solve for x: $\sqrt{3x} + 4 = x - 2$

    [A] 3

    [B] 9

    [C] 12

    [D] Both A and C

    [E] None of the above

    The answer is C
    Begin the solution process by isolating the radical: $\sqrt{3x} = x - 6$
    Then square both sides of the equation: $3x = x^2 - 12x + 36$
    Set the quadratic equation equal to zero: $0 = x^2 - 15x + 36$
    Factor: $0 = (x - 12)(x - 3)$
    The two possible solutions for x are 12 and 3, however when squaring both sides of an equation extraneous solutions can occur. The answers need to be checked in the original equation:
    $\sqrt{3 \cdot 12} = 12 - 6$ and $\sqrt{3 \cdot 3} \neq 3 - 6$
    $\sqrt{36} = 6$ $\sqrt{9} \neq -3$

2. Solve for x: $7 - 5x = 7x - 11 - 3x$

    [A] −2

    [B] 2

    [C] 0

    [D] $-1\frac{1}{2}$

    [E] $3\frac{1}{4}$

The answer is B.

First add $5x$ to both sides to get all the x terms on one side. Then $7 = 7x - 11 - 3x + 5x$. Combine all the x terms: $7 = 9x - 11$. Add 11 to both sides to get $18 = 9x$. Dividing both sides by 9, $x = 2$.

3. Find the zeroes of $f(x) = x^3 + x^2 - 14x - 24$

   [A] 4, 3

   [B] 3, –8

   [C] 7, –2, –1

   [D] 4, 3, 2

   [E] 4, –3, –2

The answer is E.

Possible rational roots of the equation $0 = x^3 + x^2 - 14x - 24$ are all the positive and negative factors of 24. By substituting into the equation, we find that -2 is a root, and therefore that $x + 2$ is a factor. By performing the long division $(x^3 + x^2 - 14x - 24)/(x + 2)$, we can find that another factor of the original equation is $x^2 - x - 12$ or $(x - 4)(x + 3)$. Therefore the zeros of the original function are -2, -3, and 4.

4. Expand $(x + 2)^3$

   [A] $6x$

   [B] $3x + 6$

   [C] $x^3 + 8$

   [D] $x^3 + 3x^2 + 8$

   [E] $x^3 + 6x^2 + 12x + 8$

The answer is E

To find the polynomial answer, either multiply the binomial by itself 3 times, or apply the cube formula $(A + B)^3 = A^3 + 3A^2B + 3AB^2 + B^3$

5. Which of the following sets of ordered pairs does not represent a function?

   [A] {(1,4), (2,5), (3,6)}

   [B] {(1,–1), (2,–2), (3,–3)}

   [C] {(3,1), (4,1), (5,1)}

   [D] {(1,3), (1,4), (1,5)}

   [E] All of the above do represent functions

The answer is D

A function cannot have more than one output value (y) for the same input (x).

6. Solve for x $\dfrac{2}{x} + \dfrac{3}{8} = \dfrac{5}{2x}$

   [A] $\dfrac{4}{3}$

   [B] $\dfrac{8}{3}$

   [C] 4

   [D] 8

   [E] 16

The answer is A

To clear the fractions, multiply both sides of the equation by the common denominator 8x. The cleared equation is $16 + 3x = 20$, $x = \dfrac{4}{3}$

7. Solve for y: $4x - 3y = 9$

   [A] $y = 4x - 9$

   [B] $y = 4x + 3$

   [C] $y = 3 - \frac{4}{3}x$

   [D] $y = \frac{4}{3}x - 3$

   [E] $y = \frac{5}{3}x$

The answer is D

Original equation: $4x - 3y = 9$, subtract 4x from both sides
$-3y = -4x + 9$, divide each term by -3
$y = \frac{4}{3}x - 3$

8. State the domain of the function $f(x) = \frac{3x - 6}{x^2 - 25}$

   [A] $x \neq 2$

   [B] $x \neq 5, -5$

   [C] $x \neq 2, -2$

   [D] $x \neq 5$

   [E] All real numbers

The answer is B

The values of 5 and −5 must be omitted from the domain of all real numbers because if x took on either of those values, the denominator of the fraction would have a value of 0, and therefore the fraction would be undefined.

9. Solve $9x^2 - 25 = 0$

[A] $\pm \dfrac{5}{3}$

[B] ± 5

[C] 0

[D] $\pm \dfrac{5i}{3}$

[E] ± 4

The answer is A

Solving the equation for x^2 yields $x^2 = \dfrac{25}{9}$. Take the square root of both sides to find that $x = \pm \dfrac{5}{3}$.

10. What is the equation of the graph shown below?

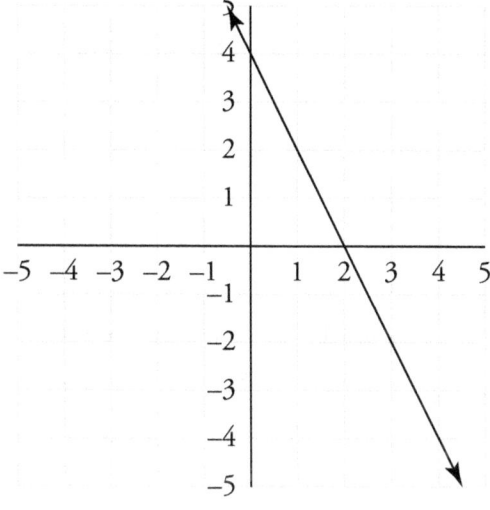

[A] $x + 2y = 4$

[B] $x - 2y = 4$

[C] $2x + y = 4$

[D] $x + 2y = -4$

[E] $x - 2y = -4$

The answer is C

When $y = 0$, solving for x yields 2 giving the point (2, 0). When $x = 0$, solving for y yields 4 giving the point (0, 4). These are the intercepts of the given graph.

11. Find the reflection of the point (4,7) over the y axis.

[A] (7,4)

[B] (4,–7)

[C] (0,7)

[D] (–4,7)

[E] (4, 0)

The answer is D

When a point (x,y) is reflected over the y axis, it converts to $(-x,y)$

12. If $Pv = nrt$, solve for n.

[A] $n = \dfrac{Pv}{rt}$

[B] $n = Pv - rt$

[C] $n = Pvrt$

[D] $n = P + v - r - t$

[E] none of the above

The answer is A

In order to isolate the variable *n*, both sides of the equation must be divided by *rt*. This results in choice A.

13. If $f(x) = 10 - 3x^2$, find $f(-5)$

    [A] −215

    [B] −65

    [C] 85

    [D] 175

    [E] 235

The answer is B

Following order of operations, $10 - 3(-5)^2 = 10 - 3(25) = 10 - 75 = -65$

14. In a class of 24 students, 3 children are randomly chosen to form a focus group. How many different groups can be made?

    [A] 8

    [B] 72

    [C] 2,024

    [D] 12,144

    [E] 13,824

The answer is C

A random selection of students for the purpose of forming a group represents a combination, as the order within the group does not matter. Therefore calculate $_{24}C_3 = 2{,}024$.

15. Given a spinner marked as shown, what is the probability of spinning a multiple of 6 or a multiple of 4?

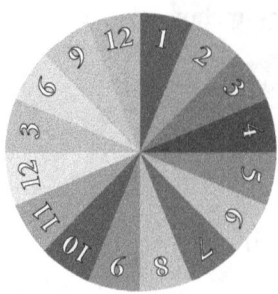

[A] $\dfrac{7}{16}$

[B] $\dfrac{5}{16}$

[C] $\dfrac{1}{2}$

[D] $\dfrac{1}{8}$

[E] $\dfrac{3}{8}$

The answer is E

Since there are 4 spaces (6 twice and 12 twice) representing multiples of 6, P(multiple of 6) = $\dfrac{4}{16}$. There are 4 spaces representing multiples of 4 (4, 8, 12,12) but only count two of them as the twelves have already been accounted for. This adjusted probability is then $\dfrac{2}{16}$. To find the chance of one case or the other, simply add the probabilities together. $\dfrac{4}{16}+\dfrac{2}{16}=\dfrac{6}{16}=\dfrac{3}{8}$

16. Given a standard deck of cards, a king is drawn and placed face up on the table. What is the probability that a second king will be randomly chosen in a second draw?

   [A] $\dfrac{1}{13}$

   [B] $\dfrac{4}{13}$

   [C] $\dfrac{3}{52}$

   [D] $\dfrac{3}{51}$

   [E] $\dfrac{1}{3}$

The answer is D

If one king has already been removed from the deck, the total number of kings remaining is 3 out of 51, the total number of cards left in the deck.

17. If P(R) = 0.6, what is the P(not R)?

   [A] 1

   [B] 0.6

   [C] 0.4

   [D] 0.3

   [E] 0

The answer is C

The events "R" and "not R" are complementary and their probabilities must add to 1.

18. If A and B are independent events, and P(A) = 20% and P(B) = 70%, find P(A ∩ B).

   [A] 90%

   [B] 50%

   [C] 35%

   [D] 14%

   [E] 5%

   The answer is D
   When two events are independent, the chance that they will both occur is the product of their independent probabilities.

19. A spinner contains sections for the numbers 1–12 but the multiples of three appear twice. What is the expected value of a spin?

   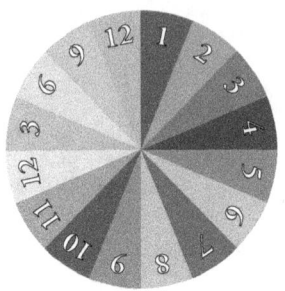

   [A] 1

   [B] 3

   [C] 6

   [D] 6.75

   [E] 10.5

The answer is D

To calculate expected value, multiply the value of each outcome by its probability and add the products together.

20. In the first four games of the season, a basketball team scores 25, 30, 33, and 28 points per game. How many points does the team need to score in their 5th game to reach an average of 30 points per game?

    [A] 29

    [B] 30

    [C] 32

    [D] 33

    [E] 34

The answer is E

Set up an equation to represent the calculation of the average, where x = the score of the 5th game:

$$\frac{\text{total}}{\text{\# of values}} = \frac{25+30+33+28+x}{5}$$
$$30 = \frac{116+x}{5}$$
$$150 = 116+x$$
$$x = 34$$

21. Find the mode for the given set of data. {1, 12, 22, 32, 32}

    [A] 2

    [B] 22

    [C] 27

    [D] 31

    [E] 32

The answer is E

The mode represents the piece of data that appears the most often in a set.

22. Find the range for the given set of data. {1, 12, 22, 32, 32}

    [A] 1

    [B] 22

    [C] 27

    [D] 31

    [E] 32

The answer is D

The range is defined as the difference between the maximum and minimum values in a set of data. $32 - 1 = 31$.

23. If the median test score in a class is 88%, which of the following statements correctly follow?

    [A] The teacher grades using the Normal Distribution.

    [B] The mean is also 88%.

    [C] The same number of students scored above 88% as scored below 88%.

    [D] The range of test scores is 60–100%.

    [E] None of the above.

The answer is C

The median is the number exactly in the middle of a set of data listed in numerical order. Therefore the same number of data pieces exist on each side of the median.

24. Which of the following represents a true statement?

[A] The range of a set of data represents the numerical distance between the smallest and largest pieces of data.

[B] A standard deviation is the same as a quartile.

[C] A set of data can have only one outlier

[D] A set of data can have only one mode

[E] The mean and median of a set of data are the same value.

The answer is A

Choice A describes the definition of range. For choice B, a quartile always represents one fourth of the data, but a standard deviation's representation can vary depending on the data. Choice C and D are false because a set of data can certainly have multiple outliers or modes. And in choice E, a mean and median can be the same value, but are not guaranteed to be so.

25. Which of the following is not a conclusion that can be drawn from the bar graph below?

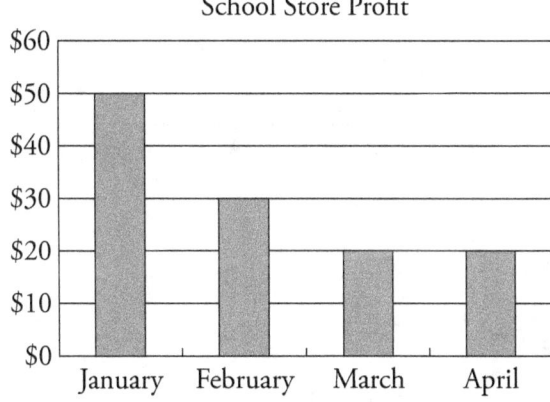

[A] Profit for December was approximately between $50 and $60.

[B] Profits for March and April were approximately equal.

[C] Profits decreased since January.

[D] The average profit is approximately $30 per month.

[E] Both A and D are invalid conclusions.

The answer is A

December is not part of the time period displayed in the graph. Furthermore, the decreasing trend present in the graph cannot predict a profit value for December.

26. The phrase that best describes this graph of data is

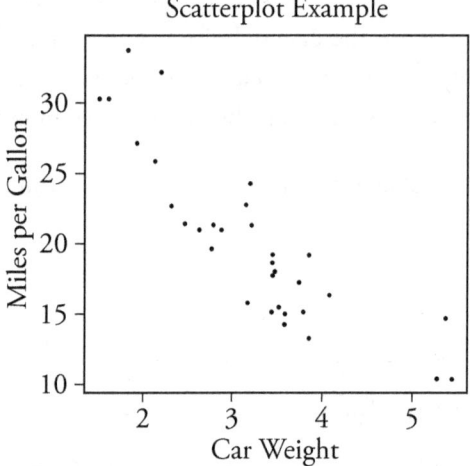

[A] Positively Correlated

[B] Negatively Correlated

[C] Exponentially Correlated

[D] Continuously Correlated

[E] Not Correlated

The answer is B

When imagining the best fit path through the points, it can be seen as a straight line with a negative slope, which describes a negative correlation.

27. Which of the following statements is true regarding the graph of a set of data that is normally distributed?

    [A] The maximum value of the graph is the same as the mean, median, and mode of the set of data.

    [B] The standard deviation is ±3

    [C] The graph crosses the x axis at the 5th standard deviation from the mean.

    [D] All of the above

    [E] None of the above

The answer is A

Choice A describes the characteristics of the normal curve. Choice B cannot be determined from the given information and choice C is not true as the graph of a normal curve lies entirely above the x axis.

28. 60 is what percent of 400?

    [A] 10%

    [B] 15%

    [C] 24%

    [D] 25%

    [E] 69%

The answer is B

Translate the given relationship into an equation, where p = percent.

$$60 = p \times 400$$
$$p = \frac{60}{400} = .15 = 15\%$$

29. A customer buys a sweater for $24, pants for $28, and socks for $4. Find the total amount the customer will be charged if the sales tax is 8.5%.

[A] $4.76

[B] $47.60

[C] $56.55

[D] $60.76

[E] $64.50

The answer is D

The subtotal before tax is 24 + 28 + 4 = 56. To calculate the after tax total, find 8.5% of the subtotal and add to the subtotal.

$$56(0.085) + 56 = 56(1.085) = 60.76$$

30. If the price of a car rises from $15,900 to $17,000, find the percent increase.

[A] 6.5%

[B] 6.9%

[C] 9.4%

[D] 11%

[E] 25%

The answer is B

To calculate percent increase, compare the difference to the original cost. The difference in price is 17,000 − 15,900 = 1,100. Calculate $\frac{1,100}{15,900} \approx 0.069 = 6.9\%$

31. A 15% service charge was added to labor costs making the final billed amount $275. How much was the cost of labor?

[A] $220

[B] $234

[C] $239

[D] $259

[E] $260

The answer is C
  Let x = the labor cost. To represent the service charge calculation, use the equation: $x + 0.15x = 275$
$$1.15x = 275$$
$$x \approx 239$$

32. If a credit card company applies an APR of 22% to its monthly statements, find the effective interest rate charged.

[A] 10%

[B] 22.2%

[C] 23.8%

[D] 24.4%

[E] 25%

The answer is D
  To calculate the effective rate, use the formula $\left(1 + \frac{r}{n}\right)^n - 1$ where r = the APR and n = the yearly compounding frequency.
  Evaluate $\left(1 + \frac{0.22}{12}\right)^{12} - 1 \approx 0.2435 \approx 24.4\%$

33. $2,000 is invested in an account paying 3.7% annual interest compounded quarterly. Find the balance of the account after 10 years.

[A] $2,037.04

[B] $2,740

[C] $2,890.55

[D] $2,900

[E] $3,010.25

The answer is C

Use the compound interest formula $P\left(1+\frac{r}{n}\right)^{nt}$ where r = the APR, n = the yearly compounding frequency, t = the total investment time, and P = the initial deposit.

Evaluate $2{,}000\left(1+\frac{0.037}{4}\right)^{4\cdot 10} \approx 2{,}890.55$

34. Find the balance, after 8 years, when $2,000 is deposited in an account with an APR of 5% compounded continuously.

[A] $2,040

[B] $2,954.91

[C] $2,983.65

[D] $2,800

[E] $2,851.23

The answer is C

When compounding continuously, use the formula Pe^{rt} where r = the APR, t = the total investment time, and P = the initial deposit.

Evaluate $2{,}000e^{0.05\cdot 8} \approx 2{,}983.65$.

35. Find the present value needed to achieve a future value, in 15 years' time, of $10,000 if the investment earns an APR of 8% compounded annually.

[A] $2,500

[B] $2,815

[C] $3,000

[D] $3,129

[E] $3,152

The answer is E

Use the compound interest formula set equal to the future value:

$$A = P\left(1 + \frac{r}{n}\right)^{nt}$$

$$10,000 = P\left(1 + \frac{.08}{1}\right)^{1 \cdot 15}$$

$$10,000 = P(1.08)^{15}$$

$$3,152 \approx P$$

36. A customer is choosing between 3 different interest bearing accounts.
    - Choice 1 offers an APR of 6% compounded quarterly.
    - Choice 2 offers an APR of 5.5% compounded monthly.
    - Choice 3 offers an APR of 5% compounded weekly.

Which will produce a higher future value over a set period of time?

[A] Choice 1

[B] Choice 2

[C] Choice 3

[D] They all offer the same future value

[E] There is not enough information given to determine

The answer is A

Use the compounding interest formula to compare the accounts.

$$A_1 = P\left(1+\frac{0.06}{4}\right)^{4t} = P\left(\left(1+\frac{0.06}{4}\right)^4\right)^t \approx P(1.061)^t$$

$$A_2 = P\left(1+\frac{0.055}{12}\right)^{12t} = P\left(\left(1+\frac{0.055}{12}\right)^{12}\right)^t \approx P(1.056)^t$$

$$A_3 = P\left(1+\frac{0.05}{52}\right)^{52t} = P\left(\left(1+\frac{0.05}{52}\right)^{52}\right)^t \approx P(1.051)^t$$

With the deposit, P, and the amount of time, t, invested remaining equivalent in all 3 scenarios, choice 1 offers the best future value as it has the greatest base for exponential growth.

37. Given $\triangle DOG$ is isosceles with $\overline{DO} \cong \overline{OG}$ and $m\angle O = 22°$, find $m\angle D$.

   [A] 180°

   [B] 158°

   [C] 79°

   [D] 22°

   [E] none of the above

The answer is C

Since $\overline{DO}$ and $\overline{OG}$ are the congruent legs of the triangle, the congruent base angles are $\angle D$ and $\angle G$. Let the measure of a base angle $= x$.

$$180 = 2x + 22, \; x = 79$$

38. Find the surface area of a box which is 3 feet wide, 5 feet tall, and 4 feet deep.

   [A] 47 sq. ft.

   [B] 60 sq. ft.

   [C] 94 sq. ft

   [D] 188 sq. ft.

   [E] 200 sq. ft.

The answer is C

Let's assume the base of the rectangular solid (box) is 3 by 4, and the height is 5. Then the surface area of the top and bottom together is 2(12) = 24. The sum of the areas of the front and back are 2(15) = 30, while the sum of the areas of the sides are 2(20) = 40. The total surface area is therefore 94 square feet.

39. Based on the diagram below

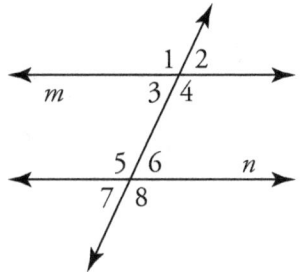

If line *m* is parallel to line *n*, then which pair of angles is congruent?

   [A] 1 and 2

   [B] 3 and 5

   [C] 2 and 8

   [D] 2 and 7

   [E] No pair of angles is congruent

The answer is D

When two parallel lines are cut by a transversal, the alternate exterior angles are congruent.

40. Which of the following is true about a parallelogram?

    [A] The opposite sides are congruent.

    [B] The diagonals are congruent.

    [C] All four angles are congruent.

    [D] The sum of the interior angles is 180°.

    [E] All of the above

The answer is A

Choices B and C are true for rectangles. Choice D is true for triangles.

41. Which theorem can be used to prove $\triangle BAT = \triangle MKT$?

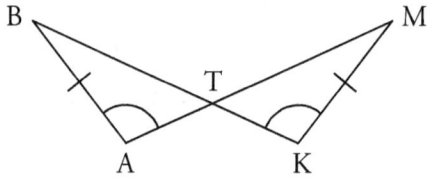

    [A] SSS

    [B] ASA

    [C] SAS

    [D] AAS

    [E] CPCTC

The answer is D

Since the angles at T are vertical angles and therefore congruent, the triangles can be proved congruent by the Angle-Angle-Side Postulate.

42. Which of the following is *not* true? An equilateral triangle:

   [A] has three equal sides

   [B] has three equal angles

   [C] has three acute angles

   [D] has three obtuse angles

   [E] has 3 vertices

The answer is D

An equilateral triangle, by definition, has three equal sides. So the three angles are equal as well and each is an acute angle equal to 60°.

43. An architect who wants to increase the area of a square room by a factor of 100 should

   [A] Make each side of the room 100 times longer

   [B] Make each side of the room 10 times longer

   [C] Double the height of the ceiling

   [D] Make the room rectangular

   [E] Make no change in the dimensions

The answer is B

The area of a square is found by squaring the length of its side. If a side of the square is originally n, and the area n^2, then making a side 10 times longer, $10n$, will result in an area of $(10n)^2 = 100n^2$ which is 100 times greater than the original area.

44. If quadrilateral *ABCD* is congruent to *PQRS*, then which of the following is true?

[A] $\overline{AD} \cong \overline{PS}$

[B] $\overline{AB} \cong \overline{QR}$

[C] $\angle A \cong \angle S$

[D] $\angle B \cong \angle P$

[B] None of the above

The answer is A

Corresponding parts of congruent figures are congruent.

45. Given A = {H, O, R, S, E} and B = {C, O, W}, find A ∩ B.

[A] {O}

[B] { H, O, R, S, E, C, O, W}

[C] {P, I, G}

[D] {H, C}

[E] ∅

The answer is A

The intersection of two sets represents the element(s) present in both sets. In this case, that is the letter "O."

46. Which statement below is true, based on the given Venn diagram?

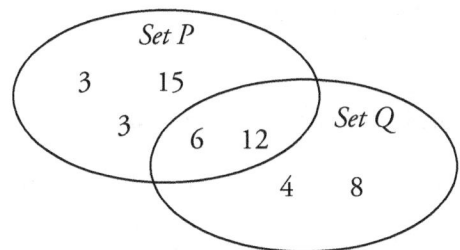

[A] P ⊂ Q

[B] P = 2Q

[C] 8 ∈ P

[D] P ∩ Q = {6, 12}

[E] None of the above

The answer is D

The numbers 6 and 12 are in both sets P and Q, thereby qualifying as the intersection.

47. Given the conditional statement: "If you heat water to 212°F, it boils," find the converse.

[A] If you chill water to 32°F, it freezes.

[B] If water is boils, it is 212°F.

[C] If you don't heat water to 212°F, then it won't boil.

[D] If you boil water, it turns to steam.

[E] None of the above.

The answer is B

The converse of a conditional is found by switching the hypothesis with the conclusion.

48. Which statement below represents a valid conclusion?

   [A] If n, then p. If p, then w. Therefore if n, then w.

   [B] If b, then q. Therefore if q, then b.

   [C] If k, then t or x. Not x. Therefore, if k, then t.

   [D] All of the above are valid.

   [E] Both A and C are valid.

The answer is E

Choice A is valid according to the Law of Syllogism. Choice C represents disjunctive syllogism. Choice B is incorrect because not all conditional statements are reversible.

49. Choose the most appropriate conclusion, based on the following statements;
    - Every rhombus is a parallelogram
    - Every square is a rhombus and a rectangle.
    - A trapezoid is not a parallelogram
    - Every parallelogram is a quadrilateral.

    Conclusion:

    [A] A parallelogram is a square.

    [B] A trapezoid is not a rhombus.

    [C] A rectangle is a square.

    [D] A parallelogram is a rhombus.

    [E] A quadrilateral is a rhombus.

The answer is B

Since a trapezoid is not a parallelogram, and a rhombus is a parallelogram, a trapezoid cannot be a rhombus.

50. Which statement below is true, based on the given Venn diagram?

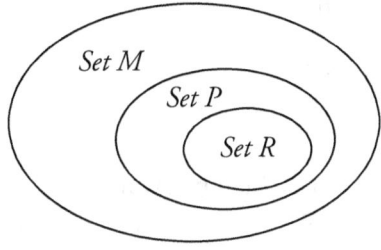

[A] R ⊂ M

[B] M ⊂ R

[C] P ∪ R = M

[D] R ∩ P = ∅

[E] None of the above

The answer is A

Since R is contained completely in M, it is a subset of M.

51. The Venn diagram below represents sports played by boys in a 6th grade class. Which statement below is true, based on the diagram?

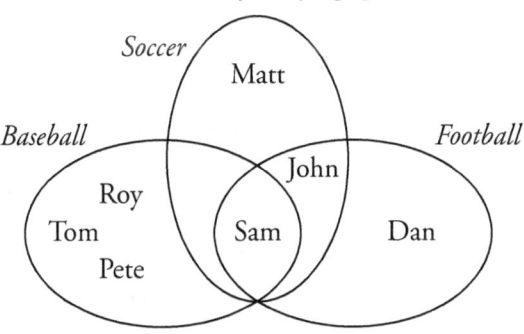

6th Grade Boys Playing Sports

Rationales for Test 2   **247**

[A] Matt plays all 3 sports.

[B] 3 boys play baseball.

[C] John plays both football and soccer.

[D] Andy plays football.

[E] None of the above.

The answer is C

Choice A is false because Matt plays only soccer. Choice B is false because a total of 4 boys play baseball. Choice C is correct as John is part of both the Soccer and Football groups. Choice D is false because there is no "Andy" included in the diagram.

52. Given M = {1, 2, 3} and P = {3, 4}, find M × P.

    [A] {1, 2, 3, 4}

    [B] {1, 4}

    [C] {(1, 3), (2, 4)}

    [D] {(1, 3), (1, 4), (2, 3), (2, 4), (3, 3), (3, 4)}

    [E] ∅

The answer is D

The Cartesian product M × P creates the set of all possible ordered pairs (m, p) such that m ∈ M and p ∈ P.

53. Which notation below represents the data graphed on the following number line?

    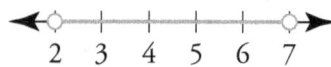

[A] {2, 7}

[B] {3, 4, 5, 6}

[C] $\{x \mid 2 < x < 7\}$

[D] $\{x \mid x > 2 \text{ and } x < 7\}$

[E] Both C and D

The answer is E

The notation shown in C and D reads "all numbers x such that..." followed by a statement of inequality that, in both cases, correctly matches the graphed data. While choice B correctly includes the integers graphed on the number line, it fails to include the infinite amount of rational values in between the whole numbers.

54. Simplify $\dfrac{|8-40|+12}{2}$

[A] 20

[B] 22

[C] 24

[D] 30

[E] 38

The answer is B

Simplify the numerator before performing the (denominator) division. Treat absolute value bars like parenthesis.

$$\dfrac{|8-40|+12}{2} = \dfrac{|-32|+12}{2} = \dfrac{32+12}{2} = \dfrac{44}{2} = 22$$

55. Which expression below is divisible by 6?

[A] $x + 6$

[B] $6x$

[C] $3x(2y)^4$

[D] All of the above

[E] B and C

The answer is E

A monomial is divisible by 6 if it has a coefficient of 6, or if it contains factors of 2 and 3. Choice A is not necessarily divisible by 6 as the 6 is not a factor of the expression.

56. The Commutative Property applies to which of the following operations?

[A] Addition

[B] Multiplication

[C] Division

[D] All of the above

[E] A and B

The answer is E

The Commutative Property is not true for division (or subtraction), but is true for addition and multiplication.

57. Amy needs 18 ft of rope. The rope costs $2.75 per yard. How much should she expect to pay?

[A] $2.75

[B] $16.50

[C] $20.75

[D] $49.50

[E] $57.75

The answer is B

Use the conversion factor of $\frac{1 \text{ yd}}{3 \text{ ft}}$: $18 \text{ ft} \left(\frac{1 \text{ yd}}{3 \text{ ft}}\right)\left(\frac{\$2.75}{1 \text{ yd}}\right) = \$16.50$

58. Which of the following is an irrational number?

[A] $\frac{1}{3}$

[B] -0.6

[C] $\sqrt{36}$

[D] $\sqrt{37}$

[E] All of the above are irrational numbers

The answer is D

An irrational number is a number that is not rational; that is, it cannot be written as a ratio. Choice B, a decimal, can be rewritten as a fraction, or ratio, as can choice C, which is really just the number 6. Since 37, in choice D, is not a perfect square, it is represented by a non-terminating decimal with no pattern and therefore cannot be rewritten as a fraction.

59. Calculate $(2.4 \times 10^8)(4.7 \times 10^{10})$

[A] 11.28×10^{80}

[B] 1.128×10^{81}

[C] 1.128×10^{19}

[D] 1.128×10^{18}

[E] 11,280,000,000

The answer is C

To calculate the scientific notation product, first multiply the leading numbers, and add the exponents on the 10's. This results in 11.28×10^{18}. Since the leading number of a scientific notation value should be between 1 and 10, remove one factor of 10 from 11.28, converting it to 1.128, and include it with the 10^{18}, which increases the exponent by one.

60. If 2 cups of chocolate chips are used in a recipe to make 60 cookies, how many cups of chocolate chips will be needed to make 210 cookies?

[A] 4 cups

[B] 7 cups

[C] 8 cups

[D] 10 cups

[E] 30 cups

The answer is B

The solution to this problem can be found by setting up and solving a proportion:

$$\frac{2 \text{ cups}}{60 \text{ cookies}} = \frac{x \text{ cups}}{210 \text{ cookies}}$$
$$60x = 2(210)$$
$$x = 7$$

Sample Test 3

Sample Test Questions

Directions: Read each item and select the best response.

1. Which choice below represents the phrase "3 less than 5 times a number?"

   [A] $3 - 5n$

   [B] $5n - 3$

   [C] $3 < 5n$

   [D] $5(n - 3)$

   [E] None of the above

2. Simplify the following expression $\dfrac{x^2 + 5x + 6}{x + 3}$

   [A] $x + 6$

   [B] $x + 3$

   [C] $x + 2$

   [D] $6x + 9$

   [E] $x^2 + 4x + 2$

3. Of the choices listed below, which one is NOT equivalent to the other four?

[A] $(4x^5y)^2$

[B] $16x^{2(5)}y^2$

[C] $2^4 x^{10} y^2$

[D] $4^2 x^{25} y^2$

[E] $\left(\dfrac{1}{4x^5y}\right)^{-2}$

4. Which choice below could correctly appear during the solving of the given equation?

$$15 + 3x = -8x$$

[A] $15 = -5x$

[B] $18 = -8x$

[C] $15 = 11x$

[D] $15 + 11x = 0$

[E] $5 + x = -5x$

5. Solve $12x^2 - 2 = 3x^2 + 23$

[A] 2.78

[B] 4

[C] ±4

[D] $\pm\sqrt{10}$

[E] $\pm\dfrac{5}{3}$

6. In the solution to the system of equations, what is the value for x?

$$\begin{cases} 5x + y = 8 \\ 3x - 4y = 14 \end{cases}$$

[A] –2

[B] 1

[C] 1.6

[D] 2

[E] $\dfrac{14}{3}$

7. Solve for x. $\dfrac{4}{x+1} = \dfrac{3}{x-7}$

[A] –1

[B] 0

[C] 7

[D] 31

[E] No solution

8. Find the solution to the inequality $25 > 7 - 2x$

[A] $x < -9$

[B] $x > -9$

[C] $x < 9$

[D] $x > 9$

[E] $x > 16$

9. Find the zeros of the function $f(x) = x^2 - 12x - 13$.

   [A] −1

   [B] 0

   [C] 12

   [D] 13

   [E] Both A and D

10. Find the domain of the function $g(x) = \dfrac{x-3}{x-5}$

    [A] All real numbers x such that $x \neq 5$

    [B] All real numbers x such that $x \neq 3$ or 5

    [C] All real numbers x such that $x \neq \dfrac{3}{5}$

    [D] All real numbers x such that $x \geq 0$

    [E] All real numbers

11. If a certain bacteria grows at a rate according to the function $b(t) = 50(1.5)^{0.3t}$ where t is measured in days and $b(0) = 50$ milligrams, find the amount of bacteria present after 4 days.

    [A] 54

    [B] 81

    [C] 178

    [D] 6,487

    [E] 20,000

12. If $h(x) = 3x - 7$, find $-h(x)$.

   [A] $-h(x) = 3x + 7$

   [B] $-h(x) = 7 - 3x$

   [C] $-h(x) = 7x - 3$

   [D] $-h(x) = -10$

   [E] None of the above

13. Solve $|2x - 5| + 2 = 45$

   [A] -19

   [B] 19

   [C] 24

   [D] Both A and C

   [E] Both B and C

14. Simplify $7\sqrt{5} + 5\sqrt{7} - \sqrt{28}$

   [A] $12\sqrt{12} + \sqrt{28}$

   [B] $12\sqrt{40}$

   [C] $35\sqrt{35} + \sqrt{28}$

   [D] $7\sqrt{5} - \sqrt{7}$

   [E] $7\sqrt{5} + 3\sqrt{7}$

15. Given the point (−3, 7), find the corresponding point that is symmetric with respect to the x axis.

   [A] (3, −7)

   [B] (−3, −7)

   [C] (3, 7)

   [D] (7, −3)

   [E] (3, 0)

16. Which choice below represents the equation of a horizontal line?

   [A] $y = 2$

   [B] $x = 2$

   [C] $x + y = 2$

   [D] $y = x$

   [E] $y = x^2$

17. Find the equation of a line that passes through the origin, and is parallel to the line $2x + 7y = 14$

   [A] $y = -\frac{2}{7}x$

   [B] $y = \frac{7}{2}x$

   [C] $y = x$

   [D] $x + y = 14$

   [E] $x + y = 0$

18. Given the equation of a parabola, $y = x^2$, which equation below represents a transformation best described as a shift of 10 units up and 3 units to the left?

 [A] $y = 3x^2 + 10$

 [B] $y = 10(x + 3)^2$

 [C] $y = (x + 3)^2 + 10$

 [D] $y = (x - 3)^2 - 10$

 [E] $y = (x - 3)^2 + 10$

19. How many unique sandwiches can be made considering choices of white or wheat bread, meat choices of turkey, ham or roast beef, and an option of butter, mayonnaise, mustard, or no condiment?

 [A] 234

 [B] 64

 [C] 24

 [D] 9

 [E] 6

20. A selection of 4 players are chosen randomly from a team of 12 to fill the positions of First Base, Second Base, Third Base and Shortstop. How many different ways can these roles be filled?

 [A] 3

 [B] 48

 [C] 495

 [D] 1,240

 [E] 11,880

21. Given a spinner with the numbers one through eight, what is the probability that you will spin an even number or a number greater than four?

    [A] 1/4

    [B] 1/2

    [C] 3/4

    [D] 1

    [E] 0

22. When rolling a standard die, what is the probability of rolling two 3's in a row?

    [A] $\frac{1}{2}$

    [B] $\frac{1}{3}$

    [C] $\frac{2}{3}$

    [D] $\frac{1}{6}$

    [E] $\frac{1}{36}$

23. If $P(A) = \frac{1}{6}$, $P(B) = \frac{1}{3}$, and $P(C) = \frac{1}{2}$, what is the expected value?

    [A] $\frac{1}{10}$

    [B] $\frac{ABC}{6}$

    [C] $\frac{A + 2B + 3C}{6}$

    [D] $6(A + B + C)$

    [E] It cannot be determined

24. If there is a 20% chance of rain, what is the chance that it will not rain?

   [A] 99%

   [B] 80%

   [C] 50%

   [D] 20%

   [E] 10%

25. When considering the data presented below, the associated equation of linear regression would most likely have a slope of

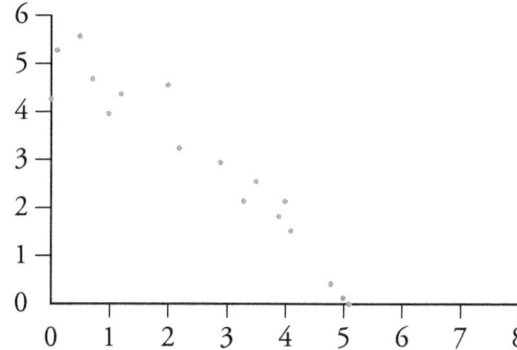

   [A] 1

   [B] –1

   [C] –2

   [D] 0

   [E] Undefined

26. Find the median of the following set of data:

    14 3 7 6 11 20

    [A] 9

    [B] 8.5

    [C] 7

    [D] 11

    [E] 6.5

27. Which of the following statements is false about the data: 2, 4, 6, 8, 10?

    [A] The mean is 6

    [B] The median is 6

    [C] The second quartile is 6

    [D] The range is 10

    [E] There is no mode

28. Faculty lunchroom data is presented below. If the school has 120 teachers and staff, how many lunches should the cafeteria expect to sell?

    Faculty and Staff Lunch Choices

    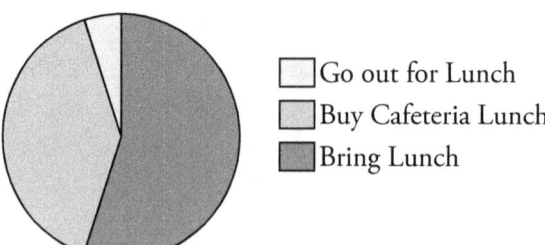

[A] 40

[B] 48

[C] 55

[D] 60

[E] 120

29. Find 65% of 210.

    [A] 115

    [B] 120

    [C] 136.5

    [D] 145

    [E] 323.1

30. 549 is 45% of what number?

    [A] 247

    [B] 594

    [C] 1110

    [D] 1220

    [E] 1549

31. At a grocery store, milk costs $3, chicken is $8, and a bag of carrots is $2. The tax rate for food is 1.9%. Find the total charge for a customer buying all 3 items and pay, including tax.

   [A] $13.19

   [B] $13.25

   [C] $14.90

   [D] $15.75

   [E] $16.00

32. Regular museum admission is $7.50. Seniors pay only $5. Find the percent savings for seniors. (Round to the nearest whole percent)

   [A] 25%

   [B] 33%

   [C] 40%

   [D] 50%

   [E] 67%

33. A landscaping company plans to increase its fees by 7%. Tree trimming currently costs $250. What will be the cost of tree trimming after the fee increase?

   [A] $257

   [B] $262.25

   [C] $267.50

   [D] $300

   [E] $357.14

34. A returning customer was given a 12% discount on his bill. If the reduced charge is $850, what would the charge have been, to the nearest dollar, without the discount?

   [A] $838

   [B] $862

   [C] $952

   [D] $957

   [E] $966

35. If an investment pays an annual interest rate of 8% compounded quarterly, find the effective annual yield.

   [A] 8.24%

   [B] 8.25%

   [C] 10%

   [D] 10.8%

   [E] 12%

36. $3,000 is invested in an account paying 2.8% annual interst compounded monthly. Find the balance of the account after 7 years.

   [A] $3,035

   [B] $3,550.25

   [C] $3,588

   [D] $3,648.75

   [E] $4,000

37. Find the balance, after 7 years, when $5,000 is deposited in an account with an APR of 4% compounded continuously.

   [A] $5,704

   [B] $6,400

   [C] $6,579.66

   [D] $6,615.65

   [E] $7,500.40

38. Find the present value needed to achieve a future value, in 10 years' time, of $15,000 if the investment earns an APR of 7% compounded annually.

   [A] $7,625

   [B] $7,800

   [C] $8,241

   [D] $8,945

   [E] $10,000

39. Which statement below correctly completes the sentence: If two lines are parallel then,

   [A] They do not intersect.

   [B] They are equidistant from each other.

   [C] They are not perpendicular.

   [D] They are in the same plane.

   [E] All of the above are true completions to the statement about parallel lines.

40. Given the diagram below, find the degree value for x.

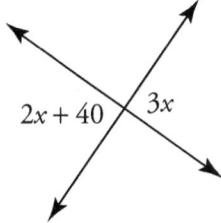

[A] 10

[B] 25

[C] 28

[D] 40

[E] 120

41. If △ABC is an acute triangle, then which of the following is true?

[A] m∠A = 60°

[B] m∠A < 60°

[C] m∠A < 90°

[D] m∠A = 90°

[E] none of the above are true

42. Given rectangle PQRS, which statement below is not necessarily true?

[A] PQRS is a parallelogram

[B] m∠P = 90°

[C] PQ = QR

[D] PQ = RS

[E] Both A and C are false

43. According to the diagram below, what is the degree value of *x*?

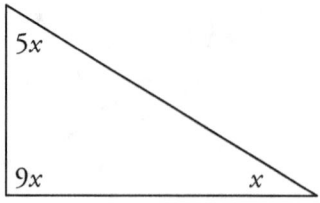

[A] 6

[B] 10

[C] 12

[D] 22.5

[E] 180

44. If △ABC~△PQR and AB = *x*, PQ = 3*x*, and the perimeter of △PQR = 24*x*, find the perimeter of △ABC.

[A] 72*x*

[B] 48*x*

[C] 20*x*

[D] 8*x*

[E] 6*x*

45. Find the length of the minor arc, $\overset{\frown}{AB}$.

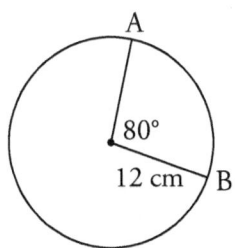

[A] 4.5cm

[B] 12cm

[C] 8π cm

[D] $\dfrac{16\pi}{3}$ cm

[E] 80°

46. Find the area of the region pictured below.

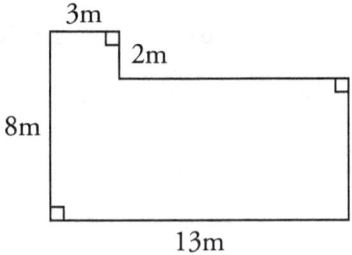

[A] 104 m²

[B] 96 m²

[C] 84 m²

[D] 80 m²

[E] 64 m²

47. A carpet square measures 16 by 16 inches, and is ¾ of an inch thick. What is the volume, in cubic inches, of a stack of 20 carpet squares?

   [A] 192

   [B] 256

   [C] 3,840

   [D] 5,120

   [E] 2,256

48. Given A = {G, R, O, U, N, D} and B = {G, R, I, N, D}, find A ∪ B.

   [A] {G, R, N, D}

   [B] {G, R, O, U, N, D, I}

   [C] {G, R, I, N, D, S}

   [D] {G, D}

   [E] ∅

49. Which statement below is true, based on the given Venn diagram?

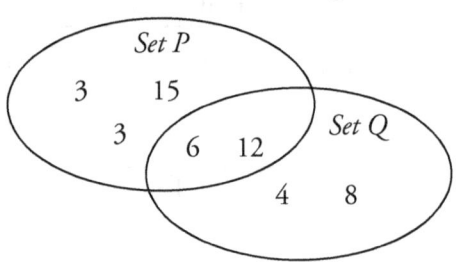

   [A] P = {3, 15, 9}

   [B] Q = {4, 6, 8, 12}

   [C] P ∩ Q = ∅

   [D] P ∪ Q = {6, 12}

   [E] None of the above

50. Given the conditional statement: "If it is raining, then the ground is wet," find the inverse.

   [A] If it is raining, then it is also thundering.

   [B] If it is not raining, then the ground is not wet.

   [C] If it is not raining today, then it will rain tomorrow.

   [D] If the ground is wet, then it is raining.

   [E] None of the above

51. Identify the true conclusion, given the following:
   $$\text{If x, then y}$$
   $$\text{If y, then x.}$$
   $$\text{If A, then x.}$$

   [A] If x, then A.

   [B] If y, then A.

   [C] If A, then y.

   [D] A cannot predict x.

   [E] A = z.

52. Which statement below is true, based on the given Venn diagram?

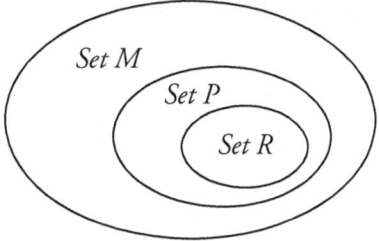

[A] R ∪ P = P

[B] R ∩ P = R

[C] P ⊂ R

[D] M ∈ R

[E] Both A and B are true.

53. The Venn diagram below represents sports played by boys in a 6th grade class. Which statement below is true, based on the diagram?

6th Grade Boys Playing Sports

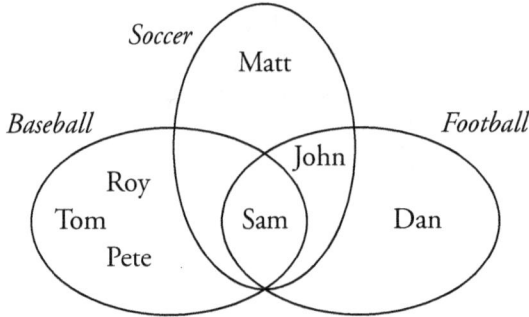

[A] 3 boys play football.

[B] 7 boys are in the class.

[C] Roy wants to play soccer.

[D] Sam and John play all 3 sports.

[E] All of the above.

54. Given E = {2, 4, 6, 8} and D = {1, 3, 5}, which of the following is not an element of E × D?

[A] (2, 8)

[B] (2, 1)

[C] (2, 5)

[D] (6, 3)

[E] (8, 1)

55. Insert mathematical symbols to make the given calculation correct.

$$3 + 5 \cdot 4 - 1 = 18$$

[A] Place parenthesis around $5 \cdot 4$

[B] Place parenthesis around $3 + 5$

[C] Place parenthesis around $4 - 1$

[D] Change the subtraction to addition of the opposite

[E] No symbols are needed. The calculation is already correct.

56. If a car travels 350 miles on 18 gallons of gas, what is its average rate of gas consumption, in miles per gallon?

[A] .05 mpg

[B] 19.4 mpg

[C] 35 mpg

[D] 194 mpg

[E] 332 mpg

57. On a certain standardized test, Student A scores 50 on math, 60 on reading, and 48 on writing. Student B scores 48 on math, 45 on reading, and 55 on writing. Which matrix below best represents this data?

[A] $\begin{bmatrix} A & 50 \\ B & 48 \end{bmatrix}$

[B] $\begin{bmatrix} 50 & 60 \\ 45 & 55 \end{bmatrix}$

[C] $\begin{bmatrix} 50 & 60 & 48 \\ 48 & 45 & 55 \end{bmatrix}$

[D] $\begin{bmatrix} 50 & 60 & 48 \\ 55 & 45 & 48 \end{bmatrix}$

[E] $\begin{bmatrix} 50 & 60 & 48 \\ 75 & 60 & 50 \\ 45 & 48 & 55 \end{bmatrix}$

58. Which of the following sets is closed under division?

    [A] Integers

    [B] Rational numbers

    [C] Natural numbers

    [D] Whole numbers

    [E] All of the above

59. Express 0.0000456 in scientific notation.

[A] 4.56×10^{-4}

[B] 45.6×10^{-6}

[C] 4.56×10^{-6}

[D] 4.56×10^{-5}

[E] 4.56×10^{5}

60. Find the GCF of $2^2 \cdot 3^2 \cdot 5$ and $2^2 \cdot 3 \cdot 7$

[A] $2^5 \cdot 3^3 \cdot 5 \cdot 7$

[B] $2 \cdot 3 \cdot 5 \cdot 7$

[C] $2^2 \cdot 3$

[D] $2^3 \cdot 3^2 \cdot 5 \cdot 7$

[E] $2 \cdot 3$

Answer Key 3

Question Number	Correct Answer	Your Answer	Question Number	Correct Answer	Your Answer
1	B		31	B	
2	C		32	B	
3	D		33	C	
4	D		34	E	
5	E		35	A	
6	D		36	D	
7	D		37	D	
8	B		38	A	
9	E		39	E	
10	A		40	D	
11	B		41	C	
12	B		42	C	
13	D		43	C	
14	E		44	D	
15	B		45	D	
16	A		46	C	
17	A		47	C	
18	C		48	B	
19	C		49	B	
20	E		50	B	
21	C		51	C	
22	E		52	E	
23	C		53	A	
24	B		54	A	
25	B		55	C	
26	A		56	B	
27	D		57	C	
28	B		58	B	
29	C		59	D	
30	D		60	C	

Rationales for Test 3

The solutions presented represent one way to find the answer to the question.

1. Which choice below represents the phrase "3 less than 5 times a number?"

    [A] $3 - 5n$

    [B] $5n - 3$

    [C] $3 < 5n$

    [D] $5(n - 3)$

    [E] None of the above

    The answer is B

    Choice B expresses the operations in the correct order. Choice C expresses 3 *is* less than 5 times a number.

2. Simplify the following expression $\dfrac{x^2 + 5x + 6}{x + 3}$

    [A] $x + 6$

    [B] $x + 3$

    [C] $x + 2$

    [D] $6x + 9$

    [E] $x^2 + 4x + 2$

    The answer is C

    Reduce the numerator and denominator by identifying their common factors.

    $$\frac{x^2 + 5x + 6}{x + 3} = \frac{(x+2)(x+3)}{(x+3)} = x + 2$$

3. Of the choices listed below, which one is NOT equivalent to the other four?

[A] $(4x^5y)^2$

[B] $16x^{2(5)}y^2$

[C] $2^4 x^{10} y^2$

[D] $4^2 x^{25} y^2$

[E] $\left(\dfrac{1}{4x^5y}\right)^{-2}$

The answer is D

Choice D is incorrect because the power of x must be 10, not 25

4. Which choice below could correctly appear during the solving of the given equation?

$$15 + 3x = -8x$$

[A] $15 = -5x$

[B] $18 = -8x$

[C] $15 = 11x$

[D] $15 + 11x = 0$

[E] $5 + x = -5x$

The answer is D

Choice D arises when 8x is added to both sides of the equation.

5. Solve $12x^2 - 2 = 3x^2 + 23$

   [A] 2.78

   [B] 4

   [C] ±4

   [D] $\pm\sqrt{10}$

   [E] $\pm\dfrac{5}{3}$

The answer is E

Start by solving for x^2:

$9x^2 = 25$

$x^2 = \dfrac{25}{9}$

Next take the plus or minus square root of both sides. This results in choice C.

6. In the solution to the system of equations, what is the value for x?

$$\begin{cases} 5x + y = 8 \\ 3x - 4y = 14 \end{cases}$$

   [A] −2

   [B] 1

   [C] 1.6

   [D] 2

   [E] $\dfrac{14}{3}$

The answer is D

Using the substitution method, solve the first equation for y: $y = 8 - 5x$ and substitute into the second equation:

$$3x - 4(8 - 5x) = 14$$
$$3x - 32 + 20x = 14$$
$$23x = 46$$
$$x = 2$$

7. Solve for x. $\dfrac{4}{x+1} = \dfrac{3}{x-7}$

   [A] −1

   [B] 0

   [C] 7

   [D] 31

   [E] No solution

The answer is D

Start the solution process with cross multiplication.

$$4(x - 7) = 3(x + 1)$$
$$4x - 28 = 3x + 3$$
$$x = 31$$

8. Find the solution to the inequality $25 > 7 - 2x$

   [A] $x < -9$

   [B] $x > -9$

   [C] $x < 9$

   [D] $x > 9$

   [E] $x > 16$

The answer is B

When dividing both sides of an inequality by a negative number, the inequality sign is reversed.

$$25 > 7 - 2x$$
$$18 > -2x$$
$$-9 < x$$
$$x > -9$$

9. Find the zeros of the function $f(x) = x^2 - 12x - 13$.

   [A] −1

   [B] 0

   [C] 12

   [D] 13

   [E] Both A and D

The answer is E

When looking at the graph of the function, the x intercepts, and therefore the zeros, are −1 and 13. Alternatively, the trinomial can be factored: $(x - 13)(x + 1)$ and show zeros of 13 and −1.

10. Find the domain of the function $g(x) = \dfrac{x - 3}{x - 5}$

    [A] All real numbers x such that $x \neq 5$

    [B] All real numbers x such that $x \neq 3$ or 5

    [C] All real numbers x such that $x \neq \dfrac{3}{5}$

    [D] All real numbers x such that $x \geq 0$

    [E] All real numbers

The answer is A

In a rational function, the denominator cannot be zero. Therefore, for the given function, x cannot have a value of 5.

11. If a certain bacteria grows at a rate according to the function $b(t) = 50(1.5)^{0.3t}$ where t is measured in days and $b(0) = 50$ milligrams, find the amount of bacteria present after 4 days.

   [A] 54

   [B] 81

   [C] 178

   [D] 6,487

   [E] 20,000

The answer is B
Evaluate the function for $t = 4$: $b(4) = 50(1.5)^{0.3(4)} \approx 50(1.6267) \approx 81$

12. If $h(x) = 3x - 7$, find $-h(x)$.

   [A] $-h(x) = 3x + 7$

   [B] $-h(x) = 7 - 3x$

   [C] $-h(x) = 7x - 3$

   [D] $-h(x) = -10$

   [E] None of the above

The answer is B
If $h(x) = 3x - 7$, then $-h(x) = -(3x - 7) = 7 - 3x$

13. Solve $|2x - 5| + 2 = 45$

   [A] -19

   [B] 19

   [C] 24

   [D] Both A and C

   [E] Both B and C

The answer is D

First isolate the absolute value expression, then set up two equations to solve.

$$|2x - 5| = 43$$
$$2x - 5 = -43 \text{ or } 2x - 5 = 43$$
$$2x = -38 \text{ or } 2x = 48$$
$$x = -19 \text{ or } x = 24$$

14. Simplify $7\sqrt{5} + 5\sqrt{7} - \sqrt{28}$

    [A] $12\sqrt{12} + \sqrt{28}$

    [B] $12\sqrt{40}$

    [C] $35\sqrt{35} + \sqrt{28}$

    [D] $7\sqrt{5} - \sqrt{7}$

    [E] $7\sqrt{5} + 3\sqrt{7}$

The answer is E

First simplify $\sqrt{28}$ by factoring out any perfect squares. $\sqrt{28} = \sqrt{4 \bullet 7} = 2\sqrt{7}$

Then combine radicals as like terms. $7\sqrt{5} + 5\sqrt{7} - 2\sqrt{7} = 7\sqrt{5} + 3\sqrt{7}$

15. Given the point (−3, 7), find the corresponding point that is symmetric with respect to the *x* axis.

    [A] (3, −7)

    [B] (−3, −7)

    [C] (3, 7)

    [D] (7, −3)

    [E] (3, 0)

The answer is B

The original point lies in the second quadrant. When this point is reflected over the x axis, its symmetric image is in the third quadrant.

16. Which choice below represents the equation of a horizontal line?

    [A] $y = 2$

    [B] $x = 2$

    [C] $x + y = 2$

    [D] $y = x$

    [E] $y = x^2$

The answer is A

A horizontal line is of the form y equals a constant.

17. Find the equation of a line that passes through the origin, and is parallel to the line $2x + 7y = 14$

    [A] $y = -\frac{2}{7}x$

    [B] $y = \frac{7}{2}x$

    [C] $y = x$

    [D] $x + y = 14$

    [E] $x + y = 0$

The answer is A

Find the slope of the given line by solving the equation for y.

$$2x + 7y = 14$$
$$7y = -2x + 14$$
$$y = \frac{-2}{7}x + 2$$

The slope of this line, shown now in slope intercept form, is $-\frac{2}{7}$. Any line parallel to this one must have the same slope. If the new line is to go through the origin, its y intercept is 0. These conditions yield the equation of the line $y = -\frac{2}{7}x + 0$, or choice A.

18. Given the equation of a parabola, $y = x^2$, which equation below represents a transformation best described as a shift of 10 units up and 3 units to the left?

    [A] $y = 3x^2 + 10$

    [B] $y = 10(x + 3)^2$

    [C] $y = (x + 3)^2 + 10$

    [D] $y = (x - 3)^2 - 10$

    [E] $y = (x - 3)^2 + 10$

The answer is C

The desired transformation moves the vertex of the parabola from its original (0, 0) location to the new point (−3, 10). The standard form of a parabola with vertex (h, k) is $y = (x - h)^2 + k$. Choice C, then, has the vertex (−3, 10)

19. How many unique sandwiches can be made considering choices of white or wheat bread, meat choices of turkey, ham or roast beef, and an option of butter, mayonnaise, mustard, or no condiment?

    [A] 234

    [B] 64

    [C] 24

    [D] 9

    [E] 6

The answer is C

Model the Fundamental Counting Principle with 3 events: 2 bread choices, 3 meat choices, and 4 possibilities for condiments. $2 \times 3 \times 4 = 24$

20. A selection of 4 players are chosen randomly from a team of 12 to fill the positions of First Base, Second Base, Third Base and Shortstop. How many different ways can these roles be filled?

[A] 3

[B] 48

[C] 495

[D] 1,240

[E] 11,880

The answer is E

Since the players are assigned to specific field locations, their selection order matters. This describes a permutation: $_{12}P_4 = 11,880$

21. Given a spinner with the numbers one through eight, what is the probability that you will spin an even number or a number greater than four?

[A] 1/4

[B] 1/2

[C] 3/4

[D] 1

[E] 0

The answer is C.

There are 6 favorable outcomes, 2, 4, 5, 6, 7, 8, and 8 possibilities. Reduce 6/8 to 3/4.

22. When rolling a standard die, what is the probability of rolling two 3's in a row?

    [A] $\frac{1}{2}$

    [B] $\frac{1}{3}$

    [C] $\frac{2}{3}$

    [D] $\frac{1}{6}$

    [E] $\frac{1}{36}$

The answer is E

Each roll of the die is an independent event. The probability of rolling a 3 is $\frac{1}{6}$. Multiply the probabilities together for a series of two independent events: $\frac{1}{6} \cdot \frac{1}{6} = \frac{1}{36}$

23. If $P(A) = \frac{1}{6}$, $P(B) = \frac{1}{3}$, and $P(C) = \frac{1}{2}$, what is the expected value?

    [A] $\frac{1}{10}$

    [B] $\frac{ABC}{6}$

    [C] $\frac{A + 2B + 3C}{6}$

    [D] $6(A + B + C)$

    [E] It cannot be determined

The answer is C

To calculate expected value, multiply the value of each outcome by its probability and add the products together.

$$\left(\frac{1}{6}\right)A + \left(\frac{1}{3}\right)B + \left(\frac{1}{2}\right)C$$

$$\frac{A}{6} + \frac{2B}{6} + \frac{3C}{6}$$

24. If there is a 20% chance of rain, what is the chance that it will not rain?

   [A] 99%

   [B] 80%

   [C] 50%

   [D] 20%

   [E] 10%

The answer is B

The events "rain" and "not rain" are complementary. Therefore their probabilities add to

25. When considering the data presented below, the associated equation of linear regression would most likely have a slope of

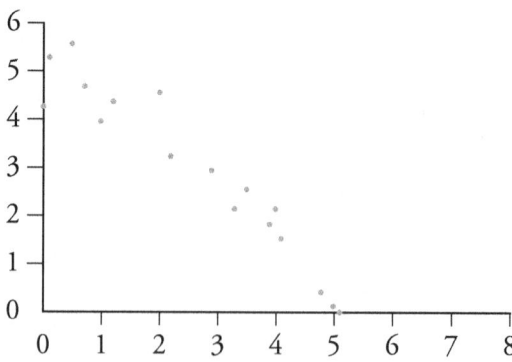

[A] 1

[B] −1

[C] −2

[D] 0

[E] Undefined

The answer is B

The equation of linear regression represents the best fit line through the points. As the points show a decreasing trend, the slope will be negative. The slope value should be close to −1 as the possible, approximate intercepts of (5, 0) and (0, 5) yield the slope calculation.

26. Find the median of the following set of data:

$$14 \ 3 \ 7 \ 6 \ 11 \ 20$$

[A] 9

[B] 8.5

[C] 7

[D] 11

[E] 6.5

The answer is A.

Place the numbers is ascending order: 3 6 7 11 14 20. Find the average of the middle two numbers: $(7 + 11)/2 = 9$.

Rationales for Test 3   **289**

27. Which of the following statements is false about the data: 2, 4, 6, 8, 10?

　[A] The mean is 6

　[B] The median is 6

　[C] The second quartile is 6

　[D] The range is 10

　[E] There is no mode

The answer is D

The second quartile of a set of data will be the same as the median, the middle number. In this case that middle is 6, which also happens to be the same as the mean:

$$\frac{2+4+6+8+10}{5} = \frac{30}{5}$$

Since none of the data values appear more than once, there is no mode.

28. Faculty lunchroom data is presented below. If the school has 120 teachers and staff, how many lunches should the cafeteria expect to sell?

Faculty and Staff Lunch Choices

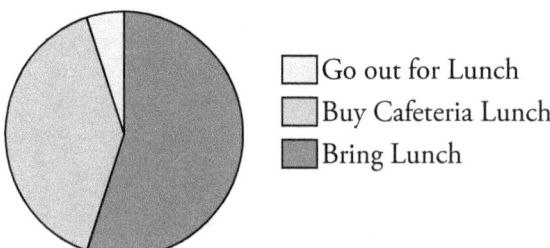

　[A] 40

　[B] 48

　[C] 55

　[D] 60

　[E] 120

The answer is B

The pie chart shows that 40% of the faculty buy a cafeteria lunch. The calculation 40% of 120 = .40 × 120 yields a result of 48.

29. Find 65% of 210.

    [A] 115

    [B] 120

    [C] 136.5

    [D] 145

    [E] 323.1

The answer is C

Translate the statement into a mathematical sentence: $0.65 \times 210 = 136.5$

30. 549 is 45% of what number?

    [A] 247

    [B] 594

    [C] 1110

    [D] 1220

    [E] 1549

The answer is D

Translate the statement into an equation, where n = the number.
$$549 = 0.45n$$
$$n = 1220$$

31. At a grocery store, milk costs $3, chicken is $8, and a bag of carrots is $2. The tax rate for food is 1.9%. Find the total charge for a customer buying all 3 items and pay, including tax.

   [A] $13.19

   [B] $13.25

   [C] $14.90

   [D] $15.75

   [E] $16.00

The answer is B

First total the groceries: $3 + 8 + 2 = 13$

Then compute the tax: $(0.019)13 = 0.25$ Adding tax to total results in choice B.

32. Regular museum admission is $7.50. Seniors pay only $5. Find the percent savings for seniors. (Round to the nearest whole percent)

   [A] 25%

   [B] 33%

   [C] 40%

   [D] 50%

   [E] 67%

The answer is B

To calculate percent change, compare the difference to the original cost. The difference in admission is $7.50 - 5 = 2.50$. Calculate $\frac{2.50}{7.50} \approx .33 = 33\%$

33. A landscaping company plans to increase its fees by 7%. Tree trimming currently costs $250. What will be the cost of tree trimming after the fee increase?

[A] $257

[B] $262.25

[C] $267.50

[D] $300

[E] $357.14

The answer is C
Find 7% of 250 and add to 250: $250(0.07) + 250 = 250(1.07) = 267.5$

34. A returning customer was given a 12% discount on his bill. If the reduced charge is $850, what would the charge have been, to the nearest dollar, without the discount?

[A] $838

[B] $862

[C] $952

[D] $957

[E] $966

The answer is E
Let $x =$ the non-discounted charge. To represent the discount calculation, use the equation: $\quad x - 0.12x = 850$
$$0.88x = 850$$
$$x \approx 966$$

35. If an investment pays an annual interest rate of 8% compounded quarterly, find the effective annual yield.

[A] 8.24%

[B] 8.25%

[C] 10%

[D] 10.8%

[E] 12%

The answer is A
To calculate the effective yield, use the formula $\left(1+\dfrac{r}{n}\right)^n - 1$ where r = the APR and n = the yearly compounding frequency.
Evaluate $\left(1+\dfrac{0.08}{4}\right)^4 - 1 \approx 0.0824 = 8.24\%$

36. $3,000 is invested in an account paying 2.8% annual interst compounded monthly. Find the balance of the account after 7 years.

[A] $3,035

[B] $3,550.25

[C] $3,588

[D] $3,648.75

[E] $4,000

The answer is D
Use the compound interest formula $P\left(1+\dfrac{r}{n}\right)^{nt}$ where r = the APR, n = the yearly compounding frequency, t = the total investment time, and P = the initial deposit.
Evaluate $3{,}000\left(1+\dfrac{0.028}{12}\right)^{12 \cdot 7} \approx 3{,}648.75$

37. Find the balance, after 7 years, when $5,000 is deposited in an account with an APR of 4% compounded continuously.

[A] $5,704

[B] $6,400

[C] $6,579.66

[D] $6,615.65

[E] $7,500.40

The answer is D

When compounding continuously, use the formula Pe^{rt} where $r =$ the APR, $t =$ the total investment time, and $P =$ the initial deposit.
Evaluate $5,000e^{0.04 \cdot 7} \approx 6,615.65$.

38. Find the present value needed to achieve a future value, in 10 years' time, of $15,000 if the investment earns an APR of 7% compounded annually.

[A] $7,625

[B] $7,800

[C] $8,241

[D] $8,945

[E] $10,000

The answer is A

Use the compound interest formula set equal to the future value:

$$A = P\left(1 + \frac{r}{n}\right)^{nt}$$

$$15,000 = P\left(1 + \frac{.07}{1}\right)^{1 \cdot 10}$$

$$15,000 = P(1.07)^{10}$$

$$7,625 \approx P$$

39. Which statement below correctly completes the sentence: If two lines are parallel then,

[A] They do not intersect.

[B] They are equidistant from each other.

[C] They are not perpendicular.

[D] They are in the same plane.

[E] All of the above are true completions to the statement about parallel lines.

The answer is E

According to the definition, parallel lines are coplanar, and maintain an equal distance apart along their entire lengths. Therefore they never intersect. Since they do not intersect, they cannot be perpendicular.

40. Given the diagram below, find the degree value for x.

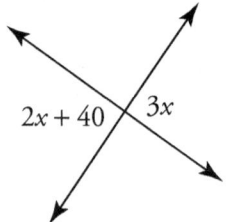

[A] 10

[B] 25

[C] 28

[D] 40

[E] 120

The answer is D

By way of the diagram, the algebraic expressions represent congruent, vertical angles. Therefore, their measures are equal.

$$3x = 2x + 40$$
$$x = 40$$

41. If $\triangle ABC$ is an acute triangle, then which of the following is true?

    [A] $m\angle A = 60°$

    [B] $m\angle A < 60°$

    [C] $m\angle A < 90°$

    [D] $m\angle A = 90°$

    [E] none of the above are true

The answer is C

By definition, all angles of an acute triangle must be less than 90 degrees.

42. Given rectangle PQRS, which statement below is not necessarily true?

    [A] PQRS is a parallelogram

    [B] $m\angle P = 90°$

    [C] PQ = QR

    [D] PQ = RS

    [E] Both A and C are false

The answer is C

Since a rectangle is a parallelogram with a right angle, choices A and B must be true. Choice D represents a pair of opposite sides so they are, in fact, of equal measure.

The sides named in choice C, however, are consecutive sides of the rectangle and are not necessarily of equal measure.

43. According to the diagram below, what is the degree value of x?

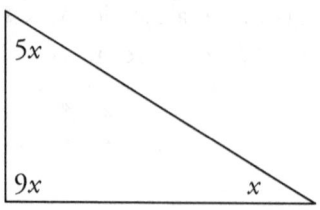

[A] 6

[B] 10

[C] 12

[D] 22.5

[E] 180

The answer is C

Since the sum of the interior angles of a triangle is 180 degrees, the following equation holds true.

$$x + 5x + 9x = 180$$
$$15x = 180$$
$$x = 12$$

44. If $\triangle ABC \sim \triangle PQR$ and $AB = x$, $PQ = 3x$, and the perimeter of $\triangle PQR = 24x$, find the perimeter of $\triangle ABC$.

[A] $72x$

[B] $48x$

[C] $20x$

[D] $8x$

[E] $6x$

The answer is D

Corresponding sides and perimeters of similar figures are proportional, so if $\frac{AB}{PQ} = \frac{x}{3x}$ then $\frac{1}{3} = \frac{\text{Perimeter of } \triangle ABC}{24x}$. The perimeter, then, must be 8x.

45. Find the length of the minor arc, $\overset{\frown}{AB}$.

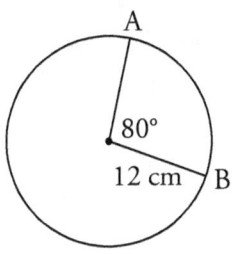

[A] 4.5cm

[B] 12cm

[C] 8π cm

[D] $\frac{16\pi}{3}$ cm

[E] 80°

The answer is D

While choice E represents the *measure* of the minor arc, the *length* of the arc is the fraction of the circumference. (Circumference of a circle = 2πr)

$$\frac{80}{360} \cdot 2\pi(12)$$

$$\frac{2}{9} \cdot 24\pi$$

46. Find the area of the region pictured below.

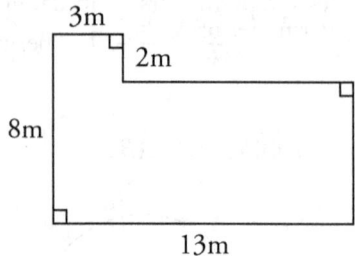

[A] 104 m²

[B] 96 m²

[C] 84 m²

[D] 80 m²

[E] 64 m²

The answer is C

Divide the space into two rectangles. Find the area of each region and add together.

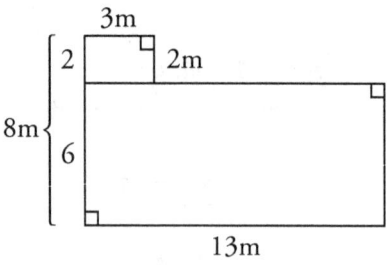

$$2(3) + 6(13) = 6 + 78 = 84$$

47. A carpet square measures 16 by 16 inches, and is ¾ of an inch thick. What is the volume, in cubic inches, of a stack of 20 carpet squares?

[A] 192

[B] 256

[C] 3,840

[D] 5,120

[E] 2,256

The answer is C

The volume of one carpet square can be calculated by multiplying its three dimensions, length, width, and height. When considering a stack of 20, the quantity becomes an additional factor in the calculation. $16 \times 16 \times 0.75 \times 20$.

48. Given A = {G, R, O, U, N, D} and B = {G, R, I, N, D}, find A ∪ B.

[A] {G, R, N, D}

[B] {G, R, O, U, N, D, I}

[C] {G, R, I, N, D, S}

[D] {G, D}

[E] ∅

The answer is B

The union of two sets combines the sets into one, without repetition.

49. Which statement below is true, based on the given Venn diagram?

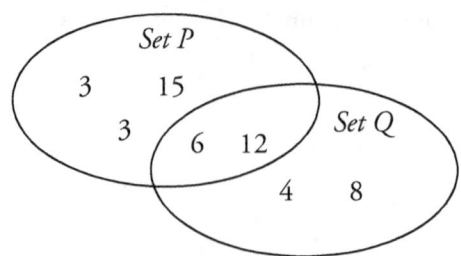

[A] P = {3, 15, 9}

[B] Q = {4, 6, 8, 12}

[C] P ∩ Q = ∅

[D] P ∪ Q = {6, 12}

[E] None of the above

The answer is B

The numbers listed show all the elements of set Q.

50. Given the conditional statement: "If it is raining, then the ground is wet," find the inverse.

[A] If it is raining, then it is also thundering.

[B] If it is not raining, then the ground is not wet.

[C] If it is not raining today, then it will rain tomorrow.

[D] If the ground is wet, then it is raining.

[E] None of the above

The answer is B

To find the inverse of a statement, take the negation of both hypothesis and conclusion.

51. Identify the true conclusion, given the following:

$$\text{If x, then y}$$
$$\text{If y, then x.}$$
$$\text{If A, then x.}$$

[A] If x, then A.

[B] If y, then A.

[C] If A, then y.

[D] A cannot predict x.

[E] A = z.

The answer is C

Choice C is the true conclusion, by the Law of Syllogism.

52. Which statement below is true, based on the given Venn diagram?

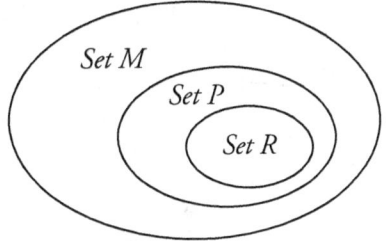

[A] R ∪ P = P

[B] R ∩ P = R

[C] P ⊂ R

[D] M ∈ R

[E] Both A and B are true.

The answer is E

Since R is a subset of P, the union of the two sets is P and the intersection of the two sets is R.

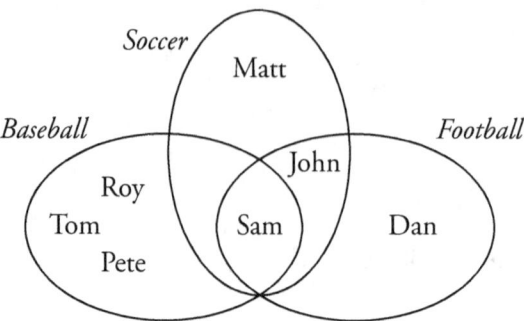

53. The Venn diagram below represents sports played by boys in a 6th grade class. Which statement below is true, based on the diagram?

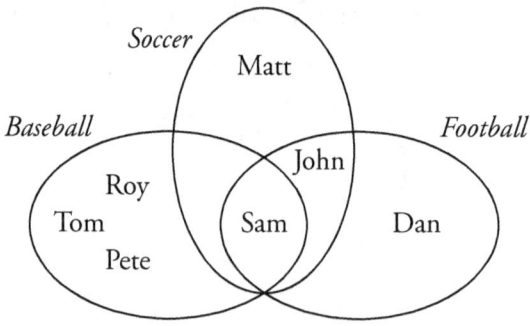

[A] 3 boys play football.

[B] 7 boys are in the class.

[C] Roy wants to play soccer.

[D] Sam and John play all 3 sports.

[E] All of the above.

304   CLEP College Level Math

The answer is A

Sam, John, and Dan are the 3 football players. Additionally, the other choices are false. 7 boys are represented in the diagram, but only sport playing boys are included. The class could have more boys that do not play sports. Roy does not play soccer, but no conclusion can be made about his desire for the sport. Only Sam plays all 3 sports.

54. Given E = {2, 4, 6, 8} and D = {1, 3, 5}, which of the following is not an element of E × D?

    [A] (2, 8)

    [B] (2, 1)

    [C] (2, 5)

    [D] (6, 3)

    [E] (8, 1)

The answer is A

The set created by the Cartesian product E × D contains all possible ordered pairs (e, d) such that e ∈ E and d ∈ D. Choice A represents an ordered pair with two elements from set E only.

55. Insert mathematical symbols to make the given calculation correct.

    $$3 + 5 \cdot 4 - 1 = 18$$

    [A] Place parenthesis around 5 · 4

    [B] Place parenthesis around 3 + 5

    [C] Place parenthesis around 4 − 1

    [D] Change the subtraction to addition of the opposite

    [E] No symbols are needed. The calculation is already correct.

The answer is C

Without inserting any symbols, the problem's answer is 22 as multiplication must be completed before the addition and subtraction. Choice C turns the calculation into

$$3+5\cdot(4-1)$$
$$3+5\cdot 3$$
$$3+15$$
$$18$$

56. If a car travels 350 miles on 18 gallons of gas, what is its average rate of gas consumption, in miles per gallon?

    [A] .05 mpg

    [B] 19.4 mpg

    [C] 35 mpg

    [D] 194 mpg

    [E] 332 mpg

The answer is B

To find the rate in miles per gallon, set up the following ratio:

$$\frac{\text{miles}}{\text{gallons}} = \frac{350}{18}$$

Division results in choice B.

57. On a certain standardized test, Student A scores 50 on math, 60 on reading, and 48 on writing. Student B scores 48 on math, 45 on reading, and 55 on writing. Which matrix below best represents this data?

[A] $\begin{bmatrix} A & 50 \\ B & 48 \end{bmatrix}$

[B] $\begin{bmatrix} 50 & 60 \\ 45 & 55 \end{bmatrix}$

[C] $\begin{bmatrix} 50 & 60 & 48 \\ 48 & 45 & 55 \end{bmatrix}$

[D] $\begin{bmatrix} 50 & 60 & 48 \\ 55 & 45 & 48 \end{bmatrix}$

[E] $\begin{bmatrix} 50 & 60 & 48 \\ 75 & 60 & 50 \\ 45 & 48 & 55 \end{bmatrix}$

The answer is C

Matrix C shows all the test data, with the rows representing each student and the columns representing math, reading, and writing.

58. Which of the following sets is closed under division?

[A] Integers

[B] Rational numbers

[C] Natural numbers

[D] Whole numbers

[E] All of the above

The answer is B

In order to be closed under division, when any two members of the set are divided, the answer must be contained in the set. This is not true for integers, natural, or whole numbers as illustrated by the counter example $11/2 = 5.5$.

59. Express 0.0000456 in scientific notation.

   [A] 4.56×10^{-4}

   [B] 45.6×10^{-6}

   [C] 4.56×10^{-6}

   [D] 4.56×10^{-5}

   [E] 4.56×10^{5}

The answer is D.

In scientific notation, the decimal point belongs to the right of the 4, the first significant digit. To get from 4.56×10^{-5} back to 0.0000456, we would move the decimal point 5 places to the left.

60. Find the GCF of $2^2 \cdot 3^2 \cdot 5$ and $2^2 \cdot 3 \cdot 7$

   [A] $2^5 \cdot 3^3 \cdot 5 \cdot 7$

   [B] $2 \cdot 3 \cdot 5 \cdot 7$

   [C] $2^2 \cdot 3$

   [D] $2^3 \cdot 3^2 \cdot 5 \cdot 7$

   [E] $2 \cdot 3$

The answer is C.

Choose the number of each prime factor that is in common.

XAMonline
The CLEP Specialist
Individual Sample Tests in ebook format with full explanations

eBooks

All 33 CLEP sample tests are available as ebook downloads from retail websites such as **Amazon.com** and **Barnesandnoble.com**

American Government	9781607875130
American Literature	9781607875079
Analyzing and Interpreting Literature	9781607875086
Biology	9781607875222
Calculus	9781607875376
Chemistry	9781607875239
College Algebra	9781607875215
College Composition	9781607875109
College Composition Modular	9781607875437
College Mathematics	9781607875246
English Literature	9781607875093
Financial Accounting	9781607875383
French	9781607875123
German	9781607875369
History of the United States I	9781607875178
History of the United States II	9781607875185
Human Growth and Development	9781607875444
Humanities	9781607875147
Information Systems	9781607875390
Introduction to Educational Psychology	9781607875451
Introductory Business Law	9781607875420
Introductory Psychology	9781607875154
Introductory Sociology	9781607875352
Natural Sciences	9781607875253
Precalculus	9781607875345
Principles of Macroeconomics	9781607875406
Principles of Microeconomics	9781607875468
Principles of Marketing	9781607875475
Principles of Management	9781607875468
Social Sciences and History	9781607875161
Spanish	9781607875116
Western Civilization I	9781607875192
Western Civilization II	9781607875208

TO ORDER

 XAMonline.com or  or

XAMonline
CLEP
Full Study Guides

CLEP College Algebra
ISBN: 9781607875598
Price: $34.95

CLEP Biology
ISBN: 9781607875314
Price: $34.95

CLEP Analyzing and
Interpreting Literature
ISBN: 9781607875260
Price: $34.95

CLEP College Composition
and Modular
ISBN: 9781607875277
Price: $19.99

CLEP College Mathematics
ISBN: 9781607875321
Price: $34.95

CLEP Psychology
ISBN: 9781607875291
Price: $34.95

CLEP Spanish
ISBN: 9781607875284
Price: $34.95

TO ORDER **X** XAMonline.com or *amazon* or **BARNES & NOBLE** BOOKSELLERS

XAMonline
CLEP Subject Series
Collection by Topic
Sample Test Approach

CLEP Literature
ISBN: 9781607875833
Price: $34.95

CLEP Foreign Language
ISBN: 9781607875772
Price: $34.95

CLEP History
ISBN: 9781607875789
Price: $34.95

CLEP Sociology
ISBN: 9781607875796
Price: $34.95

CLEP Science
ISBN: 9781607875802
Price: $34.95

CLEP Mathematics
ISBN: 9781607875819
Price: $34.95

CLEP Business
ISBN: 9781607875826
Price: $34.95

TO ORDER

XAMonline.com   or   amazon   or   BARNES & NOBLE BOOKSELLERS

XAMonline

CLEP Favorites

Collection by Topic
Sample Test Approach

CLEP Five Favorites
ISBN: 9781607875765
Price: $24.95

CLEP Military Favorites
ISBN: 9781607875512
Price: $24.95

TO ORDER    XAMonline.com   or      or

www.ingramcontent.com/pod-product-compliance
Lightning Source LLC
Chambersburg PA
CBHW060944230426
43665CB00015B/2051